CALCULUS SOLUTIONS

HOW TO SUCCEED IN CALCULUS

From Essential Prerequisites to Practice Examinations

PETER SCHIAVONE
University of Alberta

Including Ten Fully Solved Practice Examinations in Introductory Calculus

Prentice Hall Canada Inc.
Scarborough, Ontario

To Linda and Francesca

Canadian Cataloguing in Publication Data
Schiavone, Peter, 1961–
Calculus solutions

ISBN 0-13-287475-x

1. Calculus – Examinations, questions, etc.
2. Calculus. I. Title.

QA303.S34 1996 515 C96-931573-2

Prentice-Hall, Inc., Upper Saddle River, New Jersey
Prentice-Hall International (UK) Limited, London
Prentice-Hall of Australia, Pty. Limited, Sydney
Prentice-Hall Hispanoamericana, S.A., Mexico City
Prentice-Hall of India Private Limited, New Delhi
Prentice-Hall of Japan, Inc., Tokyo
Simon & Schuster Asia Private Limited, Singapore
Editora Prentice-Hall do Brasil, Ltda., Rio de Janeiro

ISBN 0-13-287475-x

Acquisitions Editor: Allan Gray
Developmental Editor: Maurice Esses
Production Editor: Amber Wallace
Production Coordinator: Deborah Starks
Cover Design: Julia Hall

1 2 3 4 5 WC 00 99 98 97 96

Printed and bound in Canada

We welcome readers' comments, which can be sent by e-mail to
collegeinfo_pubcanada@prenhall.com

CONTENTS

* For students taking the 'early transcendentals' version of beginning calculus.

PART 2

MIDTERM AND FINAL PRACTICE EXAMINATIONS WITH FULLY WORKED SOLUTIONS 117

**Includes additional (optional) questions to accommodate students taking the 'early transcendentals' version of beginning calculus.

PREFACE

This publication was written for one purpose only: to show you how to succeed in a beginning calculus course. In doing so, we focus on two specific areas:

(i) Ensuring that your prerequisite mathematics is as *fluent* and *effective* as it is *required* to be - from the start your instructor will proceed on the assumption that your prerequisite is in good *working* order i.e.

- that you can quickly form and manipulate expressions involving functions
  - for example, if $f(x) = h$, what is $f(x+h)$ where h is constant ?
- that you can complete the square in a quadratic expression
- that you can simplify and /or factor algebraic expressions quickly and effectively
- that you can solve inequalities involving polynomials and rational expressions
- that you know how to use trigonometric identities
  - for example, if $\cos 2x = 2\cos^2 x - 1$, what is the corresponding formula for $\cos 3x$?

It is important to understand that throughout your beginning calculus course you will be introduced to new concepts, new ideas and new skills, all of which build on the fundamentals (i.e. your prerequisite). Your instructor will not have time to stop and review the latter. Taking time during your calculus course to reacquaint yourself with precalculus concepts is wasteful, stressful and takes time away from the important new material. The first part of this book will ensure that you are well-prepared and that you and your instructor start on a 'level playing field'.

(ii) Maximizing your performance in course examinations (midterm and final) by providing ten sample examinations in beginning calculus - five midterm examinations and five final examinations - with *fully worked, easy-to-read, teaching-style solutions* (not just answers).

Examinations invariably make up the large majority of a student's final assessment in a beginning calculus course. As a result, most students are focused on doing well in the examinations rather than understanding all of the material. Examinations in beginning calculus are largely predictable - mainly because of the limited range and repetitive nature of the topics covered. As a result the same types of questions appear over and over again. Once the most relevant and significant areas of expertise have been identified, the task reduces to working through as many relevant problems as possible i.e. problems similar to those likely to be asked in the actual examinations - preferably practice-examination questions. In doing so, students become fluent in the relevant techniques and begin to see patterns in the solutions. Eventually the solution process becomes so systematic that the actual examination becomes an anti-climax. This is how most well-prepared students succeed in calculus. There are two key ingredients in this procedure:

(i) A good supply of practice-examination questions

(ii) Full, easy-to-read solutions to every problem

The latter make the difference! It is not difficult to obtain old examinations but fully-worked, comprehensive solutions are almost never available. Even when they are available, they are usually not well-presented and often confusing. The second part of this publication contains five midterm and five final examinations in Calculus I or beginning calculus (which include optional questions for students taking the early transcendentals version of the course) with fully worked, comprehensive and easy-to-read solutions. The presentation is based on the author's many experiences as a student, college and university mathematics instructor and director of a mathematics resource centre (a unit specifically designed to assist students make a smooth transition from high-school mathematics to beginning calculus). The examinations are rated on a scale of 1 (easy) to 5 (difficult) and include information such as mark distribution and allocated time.

Accordingly, this book is divided into two main parts. Part 1 is concerned with effective preparation for a beginning calculus course and begins with a simple, multiple-choice assessment test designed to give you an indication of both your knowledge of and fluency in the basic prerequisite techniques. This is followed by a step-by-step examination/review of what you *should* know i.e. what your instructor *expects* you to know, including the relevant techniques from basic algebra, functions, polynomials and trigonometry. We include also an optional section on exponentials and logarithms for students taking beginning calculus with *early transcendentals*. The material is specific and targeted and the emphasis is on worked examples and practice to achieve fluency (to this end, there is a problem set at the end of each section - the answers to each problem can be found in Section 8 of Part 1). We recommend that you read Part 1 of this book before your calculus course begins. This will reacquaint you with the relevant techniques and help to 'warm-up' your mathematics (particularly if you've been out of school for a significant period of time). Part 1 will serve also as a valuable reference for specific formulae and techniques *during* and *after* your calculus course. To this end we have added an index of relevant formulae and techniques in Section 9 of Part 1.

Part 2 of the book begins with five practice midterm examinations. These are followed by five practice final examinations and finally detailed solutions to all ten examinations. The solutions are written more to provide teaching assistance than to furnish a set of answers. Many include the relevant theory/formulae used in arriving at the correct answer. The reason for this comes from the author's own experience as an instructor - students learn most when applying the theory to examples, particularly when they have either someone to ask or a full set of 'teaching solutions'. Study the solutions carefully. Try to mimic the steps and the reasoning behind each solution. Practice in this context will help develop a clear and logical approach to each *type* or *class* of problem.

Finally, we should mention that this publication is intended to supplement/complement rather than replace the regular offerings of a beginning calculus course. The material is designed to fill gaps which would otherwise remain unfilled and provide valuable supplementary information aimed at enhancing performance in a course which is often notorious across campuses world-wide but which is common to students of engineering, business and science-based subjects alike. Also, it is not our intention to convince readers that mathematics is useful or fun, we leave that to the many excellent textbooks on the subject. Our sole purpose is to show you how to maximize your performance in a beginning calculus course through effective preparation for both course material and examinations.

PART 1

ARE YOU READY FOR CALCULUS ?

1. DO YOU KNOW WHAT YOUR PROFESSOR ASSUMES YOU KNOW ?

From the moment you take your seat in the first lecture of a beginning calculus course, you should expect that the professor will proceed on the assumption that you have a sound, *working* prerequisite. This means that it will be assumed that you have, 'at your fingertips', the necessary ideas and techniques from precalculus. Remember, mathematics, to a greater extent than most disciplines, is a cumulative subject: the understanding of one part depends heavily on the understanding of previous parts - and the professor usually does not have time to undertake any significant review.

As a mathematics instructor, I have found that the single most common reason as to why students find calculus particularly difficult is a lack of adequate preparation. Even more significant is the fact that many students are not even aware that they are ill-prepared! Many are lulled into a false sense of security with high averages in prerequisite (usually high school) mathematics courses. It is well-known that an excellent performance at the high school level does not necessarily translate into a similar performance at the college or university level. This is evident in beginning calculus courses. One of the main reasons for this is what I prefer to call the *'gap'*:

At the start of a beginning calculus course, what a student knows and what a professor assumes a student knows, are usually two different things.

The transition from prerequisite-level (usually high school-level) mathematics to a beginning calculus course is often discontinuous. The *gap* is a measure of this discontinuity in terms of exposure to relevant techniques and sufficient practice leading to the expected fluency in these techniques. It is important to remember that published prerequisites are necessary but not always sufficient to ensure a smooth transition to the next level. Hence, high averages in mathematics courses at the high school level are not in themselves sufficient to ensure the correct level of preparation for beginning calculus.

In Part 1 of this publication, we fill the *gap* by telling you exactly what you should know (i.e. what is assumed you know) when you start a beginning calculus course - as only a calculus professor can. We avoid much of the material often presented in precalculus courses choosing instead to concentrate only on ideas and techniques relevant to beginning calculus - precalculus in the truest sense! Consequently, the presentation is short and to the point - and you can prepare yourself in your own time, at your own pace (this avoids the added burden of a lengthy precalculus course - which is often offered in parallel to the calculus course itself!).

We begin with a simple, multiple-choice, assessment test designed to give you an indication of both your knowledge of and fluency in the basic prerequisite techniques. This is followed by a step-by-step examination/review of what you *should* know i.e. the relevant techniques from basic algebra, functions and graphs, polynomial and rational functions and trigonometry. We include also an optional section on exponentials and logarithms for students taking beginning calculus with *early transcendentals.* Throughout the material, the emphasis is on **worked examples** and there is a problem set at the end of each section. These problems are chosen specifically to develop *fluency* and thus form an integral part of the text. Try to think of mathematics as a language. Calculus can be likened to a performance in that language (e.g. delivering a speech) - unless the basic grammar and vocabulary of the language are second nature, you cannot hope to perform to the best of your ability. In calculus, the prerequisite techniques from algebra, function theory and trigonometry form the basis of "grammar and vocabulary". To maximise your performance, you must have these techniques "at your fingertips" i.e. you must achieve *fluency!* Do not fall into the trap of thinking that because you understand something, you can do it - certainly, you would not apply the same logic to learning to swim or drive a car ! Clearly, *practice* is essential. The problem sets are designed to give you the required practice, as quickly and efficiently as possible.

You should read a particular section and then try the associated problem set mimicing the procedures used in the worked examples. Remember, the key is *fluency and familiarity* - so work through the problems until you become comfortable with the particular technique and then move on to the next one. You can also use this material as a reference to 'brush-up' on a particular area once your calculus course is underway. To this end, we have included the Index of Methods and Formulae in Section 9.

2. ASSESSMENT TEST

The following test is designed to see if you are prepared for beginning calculus. Attempt all 25 questions. You should score somewhere in the range 20 - 25 (i.e. at least 80%) in approximately 40 minutes to consider yourself adequately prepared. Use the material that follows to strengthen your skills and/or to acquaint yourself with techniques which may have been omitted from your prerequisite. Re-test yourself when you feel comfortable with the problem sets offered at the end of each section. You will find the answers to the test in Section 8.

1. If $f(x) = \frac{1}{2}$ then $f(x+2h) =$

(a) $\frac{1}{2}(x+2h)$ (b) $2h$ (c) $\frac{1}{2}(x+2h)$ (d) $\frac{1}{2}$

2. The expression $f(x) = 2x^2 + x + 1, x \in R$ is

(a) Always negative (b) Always positive
(c) Sometimes negative and sometimes positive (d) Always zero

3. The solutions of the equation $x^2 + 6x + 5 = -3$ are

(a) $-5, -3$ (b) $-5, -1$ (c) $4, 2$ (d) $-4, -2$

4. If $\dfrac{-3}{x+1} - \dfrac{2}{x-4} < 0$, then

(a) $x \in (-1, 2)$ (b) $x \in (-1, 2) \cup (4, \infty)$ (c) $x \in (4, \infty)$
(d) $x \in (-\infty, -1) \cup (2, 4)$

5. The expression $\cot x \sin^2 x \tan x \csc^2 x$ simplifies to

(a) $\tan x$ (b) 1 (c) $\sin x$ (d) $\sec x \tan x$

6. Let $f(x) = \sqrt{2x^2 - 1}$. $f(\sqrt{x})$ is equal to

(a) $\sqrt{2x - 1}$ (b) $2x^2 - 1$ (c) $\sqrt{\sqrt{x} - 1}$ (d) 0

7. Rationalizing the denominator in $\dfrac{1}{2-\sqrt{3}}$ leads to

(a) $\dfrac{1}{2+\sqrt{3}}$ (b) $\sqrt{3}-2$ (c) $2+\sqrt{3}$ (d) $\dfrac{2+\sqrt{3}}{4-\sqrt{3}}$

8. $x^2 < 8x + 9$ is equivalent to

(a) $-1 < x < 9$ (b) $x < -1$ or $x > 9$ (c) $x < 1$ or $x > 8$ (d) $0 < x < 9$

9. Simplify the expression $\dfrac{\sqrt{16x^2y^2}}{(9x^2)^{-\frac{3}{2}}}$, $x, y > 0$.

(a) $\dfrac{4x^4y}{27}$ (b) $\dfrac{4y}{27x^3}$ (c) $108x^5y$ (d) $108x^4y$

10. If $x < 4$, $|x-6| + |2x-8| =$

(a) $x+2$ (b) $3x-2$ (c) $4-x$ (d) $14-3x$

11. The function $f(x) = \dfrac{1}{\sqrt{x^2+x-2}}$ has domain

(a) $x \in (-\infty, -2) \cup (1, \infty)$ (b) $x \in (-\infty, -2] \cup [1, \infty)$ (c) $x \in (-2, 1)$
(d) $x \in [-2, 1]$

12. The curves represented by the equations $y = x^2 - 2$ and $y = x$ intersect at how many distinct points?

(a) 2 (b) 1 (c) 0 (d) 3

13. $\left(\dfrac{y^2}{y+x^2}\right)\left(\dfrac{x}{y^2}+1\right)$, $x, y > 0$, can be simplified to

(a) $\dfrac{1}{y^2} + \dfrac{1}{x}$ (b) $\dfrac{1}{xy}$ (c) $\dfrac{x+y^2}{y+x^2}$ (d) 1

14. $\sqrt{1-\sin^2\theta}$ equals

(a) $\cos\theta$ (b) $\sin\theta$ (c) $|\cos\theta|$ (d) 1

15. If $f(x) = 1+\sqrt{x+1}$ and $g(x) = 3x-4$, $f(g(x)) =$

(a) $3x-4+\sqrt{4x-2}$ (b) $1+\sqrt{3(x-1)}$ (c) $1-\sqrt{3x-1}$ (d) $-3+\sqrt{x-1}$

16. The graph of $y = f(x)$ has been obtained from the graph of $y = x^4$ by displacing it one unit down and two units to the right. The formula for the function $y = f(x)$ is

(a) $y = (x-2)^4 - 1$
(b) $y = (x+2)^4 - 1$
(c) $y = (x-1)^4 - 2$
(d) $y = (x-2)^4 + 1$

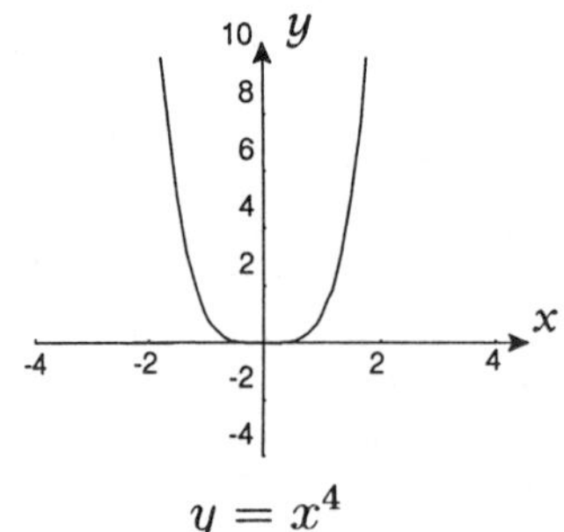

$y = x^4$

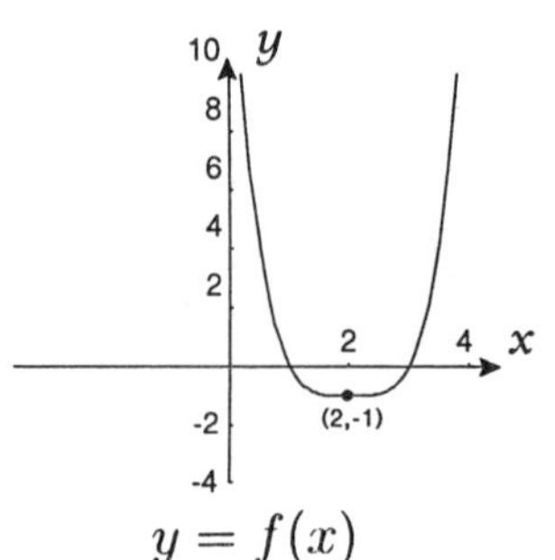

$y = f(x)$

17. The expression $4(2x+1)^{\frac{4}{5}}(2x-1)^{-\frac{1}{5}} - 8(2x-1)^{\frac{4}{5}}(2x+1)^{-\frac{1}{5}}$ can be factored to become

(a) $-4(4x^2-1)^{-\frac{1}{5}}(3-2x)$ (b) $4(4x^2-1)^{-\frac{1}{5}}(3-2x)$
(c) $8(2x-1)^{-\frac{1}{5}}(2x+3)^{-\frac{4}{5}}$ (d) $8(4x^2-1)^{-\frac{1}{5}}$

18. If $f(x) = \dfrac{2x-2}{x^2-1}$ then if $a \neq 0$, $f(2a+1) =$

(a) $\dfrac{4a-2}{4a^2-1}$ (b) $\dfrac{2a}{(2a+1)}$ (c) $\dfrac{4a+2}{(2a+1)^2}$ (d) $\dfrac{1}{a+1}$

19. If $y, w \neq 0$, the expression $\left(\dfrac{2x^3y^4}{xw^{-3}}\right)^4$ can be written as

(a) $16w^7x^6y^8$ (b) $\dfrac{2x^8y^{16}}{w^{12}}$ (c) $16x^8y^{16}w^{12}$ (d) $16x^8y^{16}w^3$

20. The expression $\dfrac{\sin 4\theta}{\cos\theta\cos 2\theta}$ simplifies to

(a) $2\sin\theta$ (b) $\dfrac{4\sin\theta}{\cos 2\theta}$ (c) $\dfrac{4\tan\theta}{\cos 2\theta}$ (d) $4\sin\theta$

21. The straight-line distance between the points $A(-1,2)$ and $B(3,-4)$ is

(a) $\sqrt{10}$ (b) $\sqrt{20}$ (c) $2\sqrt{13}$ (d) 23

22. If $f(x) = \dfrac{1}{x+1}$, the expression $\dfrac{f(x+h)-f(x)}{h}$, $h \neq 0$, is equal to

(a) $\dfrac{1}{(x+1)(x+h+1)}$ (b) $-\dfrac{1}{(x+1)(x+h+1)}$

(c) $\dfrac{-h}{(x+1)(x+h+1)}$ (d) $-\dfrac{1}{(x+1)^2}$

23. The equation of the line passing through the points $(1,5)$ and $(-1,4)$ is

(a) $2y - x - 9 = 0$ (b) $2x + 3y = 23$ (c) $2y + x = 9$
(d) $2x - 3y = 19$

24. If $f(x) = 2x^2 + 1$ and $g(x) = x + 2$ then $(g \circ f)(x) = g(f(x))$ is

(a) $2x^2 + 5$ (b) $2x^2 + 4x + 5$ (c) $2x^2 + 3$ (d) $2x^2 - 3$

25. If $\sin t = \dfrac{1}{5}$, $0 < t < \dfrac{\pi}{2}$, $\sin 2t$ is equal to

(a) $-\dfrac{2}{5}$ (b) $\dfrac{2}{5}$ (c) $\dfrac{\sqrt{24}}{25}$ (d) $\dfrac{4\sqrt{6}}{25}$

3. BASIC ALGEBRA

It is almost always the case that expressions have to be *prepared* before we can apply standard techniques from calculus (and simplified following the application of these techniques - in order to draw relevant conclusions). This requires *fluent* and *effective* factoring and simplification skills. In what follows, we present the relevant material from basic algebra.

3.1. NOTATION - SOME BASIC MATHEMATICAL VOCABULARY

We begin by reviewing some of the less frequently recognized mathematical symbols used in a beginning calculus course.

$\Rightarrow$ **implication**

$a \Rightarrow b$ can be read as " a implies b " OR " if a then b ". For example,

$$x = 2 \Rightarrow x^2 = 4$$

$\Leftrightarrow$ **equivalence**

$a \Leftrightarrow b$ can be read as " a is equivalent to b " OR " a if and only if b ". For example,

$$x^2 = 4 \Leftrightarrow x = \pm 2$$

means that the equation $x^2 = 4$ is equivalent to the equation $x = \pm 2$. That is,

$$x^2 = 4 \Rightarrow x = \pm 2 \quad \textit{and} \quad x = \pm 2 \Rightarrow x^2 = 4$$

This symbol can also be thought of as an 'equals sign' for equations (mathematical phrases).

$\in$ **belongs to**

$x \in R$ can be read as " x belongs to the set of real numbers R ".

$\exists$ **there exists or there is**

For example,

$$\exists x \in R \text{ such that } x^2 = 4$$

This can be read as " there exists or there is an $x \in R$ such that $x^2 = 4$ ".

$\forall$ **for all or for every**
For example,

$$x^2 + 1 > 0 \; , \forall x \in R$$

This can be read as " $x^2 + 1 > 0$ for every x in the set of real numbers R."

This list is by no means exhaustive! Instead, it highlights the symbols which are often assumed to be standard but with which students have traditionally lacked familiarity. We will introduce further notation throughout the text as required.

3.2. NUMBER SYSTEMS

Calculus is based on the real number system. We start with:

N: Natural numbers $\{1, 2, 3, 4, ...\}$

Z: Integers $\{... - 3, -2, -1, 0, 1, 2, 3, ...\}$

Q: Rational Numbers: a number that can be written as the quotient of two integers. That is,

$$r \in Q \Leftrightarrow r = \frac{m}{n} \; , m, n \in Z \; , n \neq 0$$

Any repeating or terminating decimal is a rational number. For example,

$$\frac{1}{2}, \quad 0 = \frac{0}{1}, \quad 5 = \frac{5}{1}, \quad \frac{-123}{452}, \quad 0.822 = \frac{822}{1000}$$

Irrational Numbers: none of the above i.e. non-terminating and non-repeating decimals are irrational.
For example,

$$\begin{aligned} & \sqrt{2}, \sqrt{10}, -\sqrt{5} \\ \pi &= 3.14159265358979..., e = 2.71828182845... \\ & 1.01001000100001... \end{aligned}$$

R: Real Numbers - all of the above i.e. the union of the sets of rational and irrational numbers.

It is interesting to note the juxtaposition of the various sets in a diagram:

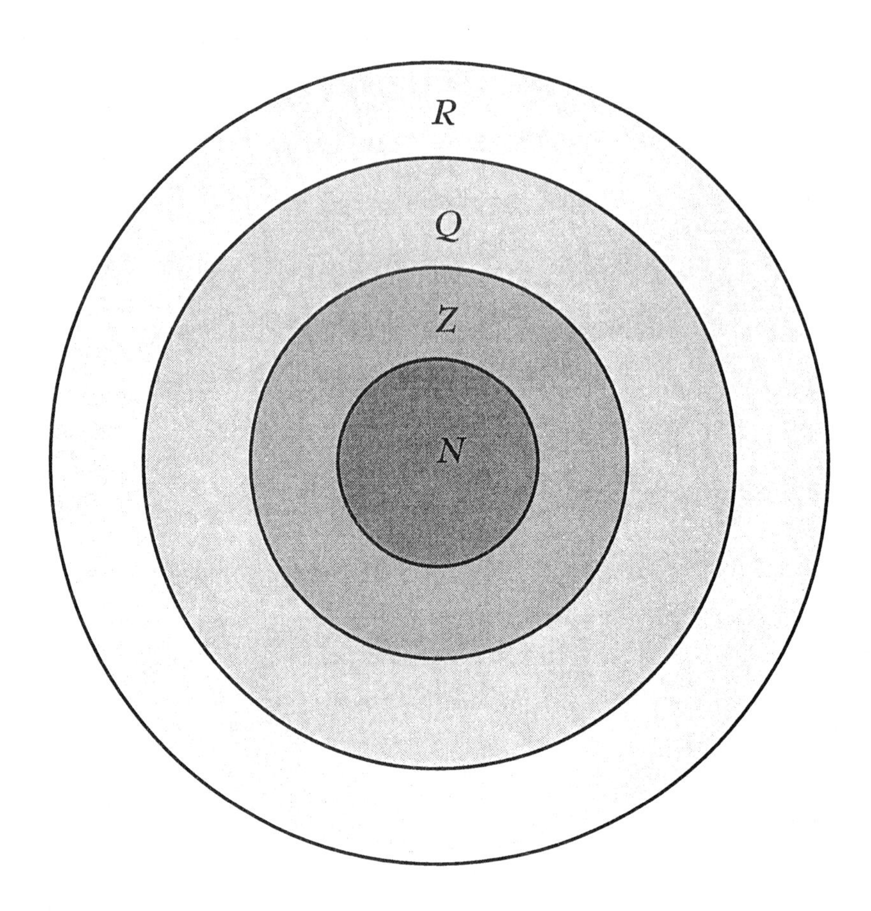

Properties of Real Numbers

Let $a, b, c, d \in R$.

(i) $-(-a) = a, \quad -(+a) = -a, \quad (-a)(-b) = ab, \quad (-a) + (-b) = -(a+b)$

(ii) $\dfrac{a}{b} = \dfrac{c}{d} \Leftrightarrow ad = bc, \quad \dfrac{a}{b} + \dfrac{c}{b} = \dfrac{a+c}{b}, \quad \dfrac{a}{b} \cdot \dfrac{c}{d} = \dfrac{ac}{bd}, \quad \dfrac{ad}{bd} = \dfrac{a}{b}, \quad b, d \neq 0$

(iii) $\dfrac{a}{b} + \dfrac{c}{d} = \dfrac{ad+cb}{bd}, \quad b, d \neq 0$ (taking a common denominator)

Warnings:

(a) Division by zero is undefined. For example, $\dfrac{3}{0}, \dfrac{-1}{0}$ are meaningless - but

$$\frac{0}{2} = \frac{0}{-1} = \frac{0}{\frac{1}{2}} = 0$$

(b)

$$\frac{a}{b+d} \neq \frac{a}{b} + \frac{a}{d} \quad \text{e.g.} \quad \frac{1}{2+4} \neq \frac{1}{2} + \frac{1}{4}$$

As simple as they may seem, these are among the two most common mistakes students bring into beginning calculus!

3.3. EXPONENTS AND ROOTS

PROPERTIES OF INTEGRAL EXPONENTS

Let $a, b \in R$, $a \neq 0$ and $m, n \in Z$.

(i) If $n \in N$, then, $a^n = \underbrace{a \cdot a \cdot a \ldots a \cdot a \cdot a}_{n \text{ times}}$ e.g. $\left(\frac{1}{2}\right)^3 = \frac{1}{2} \cdot \frac{1}{2} \cdot \frac{1}{2} = \frac{1}{8}$

(ii) $a^{-n} = \dfrac{1}{a^n}$ e.g. $3^{-2} = \dfrac{1}{3^2} = \dfrac{1}{9}$, $\left(\dfrac{1}{2}\right)^{-2} = \dfrac{1}{\left(\dfrac{1}{2}\right)^2} = \dfrac{1}{\left(\dfrac{1}{4}\right)} = 4$

(iii) $a^m a^n = a^{m+n}$ e.g. $2^3 \cdot 2^{10} = 2^{13}$

(iv) $\dfrac{a^m}{a^n} = a^{m-n}$ e.g. $\dfrac{9^3}{9^2} = 9^1 = 9$

(v) $a^0 = 1$ e.g. $9^0 = 1$

(vi) $(ab)^n = a^n b^n$ e.g. $(3 \cdot 7)^4 = 3^4 \cdot 7^4 = 194481$

(vii) $\left(\dfrac{a}{b}\right)^n = \dfrac{a^n}{b^n}, b \neq 0,$ e.g. $\left(\dfrac{5}{2}\right)^2 = \dfrac{5^2}{2^2} = \dfrac{25}{4}$

(viii) $(a^m)^n = a^{mn},$ e.g. $(2^4)^6 = 2^{24}$

Consequently,

$$\begin{aligned} a^{-1} &= \frac{1}{a}, \; a \neq 0 \\ \left(\frac{a}{b}\right)^{-1} &= \frac{1}{\left(\frac{a}{b}\right)} = \frac{b}{a}, \; a, b \neq 0 \end{aligned}$$

Warnings!
(a) 0^0 is undefined

(b)

$$(a+b)^{-1} = \frac{1}{a+b} \neq \frac{1}{a} + \frac{1}{b} = a^{-1} + b^{-1}$$

In fact, it is worth noting at this stage that more generally,

$$(a+b)^n \neq a^n + b^n$$

For example, $(x+y)^2 \neq x^2 + y^2$ (see §3.5)

We now illustrate the above rules with a few examples. Note the systematic approach of *identification* followed by *appropriate action.*

Example 3.3.1
Notice in the following examples that we work from the *inside out* i.e. we simplify the expression inside the bracket first and then apply the rules of exponents.

(i) $\left(\frac{8}{4}\right)^{12} = 2^{12} = 4096$

(ii) $(8)^{16} = (2^3)^{16} = 2^{48}$

(iii) $(2x^3y)(4y^4x) = 2 \cdot 4 \cdot x^3 \cdot x \cdot y \cdot y^4 = 8x^4y^5$ (notice how we group *like-terms*)

(iv) $(6a^3b^2)^{-5} = 6^{-5} \cdot (a^3)^{-5} \cdot (b^2)^{-5} = 6^{-5} \cdot a^{-15} \cdot b^{-10} = \frac{1}{6^5a^{15}b^{10}}$

(v) $\left(\frac{a}{b}\right)^{-n} = \left(\left(\frac{a}{b}\right)^{-1}\right)^n = \left(\frac{b}{a}\right)^n = \frac{b^n}{a^n}$

The concept of *root* arises when the exponent or power is a non-integer.

Definition 3.3.2
$a^{\frac{1}{n}} = \sqrt[n]{a}$ is called the nth root of a, $a \in R$. For example,

$$\begin{aligned} n &= 2,\ a^{\frac{1}{2}} = \sqrt{a}: \text{ square root of } a \\ n &= 3,\ a^{\frac{1}{3}} = \sqrt[3]{a}: \text{ cube root of } a \end{aligned}$$

Example 3.3.3

(a) $8^{\frac{1}{3}} = \sqrt[3]{8} = 2$

(b) $(-8)^{\frac{1}{3}} = \sqrt[3]{-8} = -2$

Warning!

$(-4)^{\frac{1}{2}} = \sqrt{-4}$ is not defined in the set of real numbers (to accommodate such quantities we have to extend the set of real numbers to the set of complex numbers - see §3.7). In general, if n is even and $a < 0$, then $\sqrt[n]{a} = a^{\frac{1}{n}}$ is not a real number. For example,

$$(-3.1456)^{\frac{1}{4}}, \quad \sqrt{-5}, \quad \left(-1.65 \times 10^3\right)^{\frac{1}{2}}$$

are not real numbers.

Hence, when working in the set of real numbers, remember that you must *never:*

(i) *Divide by zero*

(ii) *Take the even root of a negative number*

Note 3.3.4

Every positive number a, has two square roots: one positive and one negative.

$$\begin{aligned} a^{\frac{1}{2}} &= \sqrt{a} \text{ means the } \textit{positive} \text{ (principal) square root of } a \\ -a^{\frac{1}{2}} &= -\sqrt{a} \text{ means the } \textit{negative} \text{ square root of } a \end{aligned}$$

For example, $\sqrt{1} = 1$, $\sqrt{9} = 3$. The same rule applies to all *even* roots.

(b) Every real number has *exactly one* real cube root.

For example, $(27)^{\frac{1}{3}} = 3$, $(-64)^{\frac{1}{3}} = -4$. The same rule applies to all *odd* roots.

Definition 3.3.5: Rational Exponents

Let $m, n \in Z$, $n > 0$, $a \neq 0$ and $a^{\frac{1}{n}} \in R$. Then

$$a^{\frac{m}{n}} = \left(a^{\frac{1}{n}}\right)^m = \left(\sqrt[n]{a}\right)^m = \sqrt[n]{a^m}$$

The properties of integral exponents given earlier are true also when the integral exponents m and n are replaced by the rational exponents p and q provided that, whenever they appear, a^p and a^q are both real numbers.

Example 3.3.6

(i) $8^{\frac{2}{3}} = \left(8^{\frac{1}{3}}\right)^2 = \left(\sqrt[3]{8}\right)^2 = 2^2 = 4$

(ii) $(-8)^{\frac{5}{3}} = \left((-8)^{\frac{1}{3}}\right)^5 = (-2)^5 = (-1)^5 \cdot 2^5 = -32$

(iii) $(4)^{\frac{1}{2}} = \sqrt{4} = 2$

(iv) $(16)^{-\frac{5}{2}} = (16^{\frac{1}{2}})^{-5} = (\sqrt{16})^{-5} = 4^{-5} = \dfrac{1}{4^5}$

Radicals ($\sqrt{\cdot}$, $\sqrt[n]{\ }$ etc are referred to as *radical signs*)

(a) $\sqrt{a} \cdot \sqrt{b} = \sqrt{ab}$, provided $a, b \geq 0$

(b) $\dfrac{\sqrt{a}}{\sqrt{b}} = \sqrt{\dfrac{a}{b}}$, provided $a \geq 0$, $b > 0$

We can use the above rules for radicals to simplify expressions involving square roots:

Example 3.3.7

(i) $\sqrt{32} = \sqrt{16 \cdot 2} = \sqrt{16} \cdot \sqrt{2} = 4\sqrt{2}$

(ii) $\dfrac{3}{\sqrt{6}} = \dfrac{3}{\sqrt{6}} \cdot \underbrace{\dfrac{\sqrt{6}}{\sqrt{6}}}_{=1} = \dfrac{3\sqrt{6}}{6} = \dfrac{\sqrt{6}}{2}$

This last part illustrates a technique known as *Rationalizing the Denominator* i.e. removing the radical from the denominator. We do this by multiplying the original expression by **one** (so that its value remains unchanged) but in such a way as to remove the radical. This is a very common technique in beginning calculus - used mainly to simplify expressions before operating on them with a calculus-type operation (see §4.3).

(iii) $(2^{-1} + 3^{-1})^{-2} \underset{\text{working from inside out}}{=} \left(\dfrac{1}{2} + \dfrac{1}{3}\right)^{-2} = \left(\dfrac{5}{6}\right)^{-2} = \left(\dfrac{6}{5}\right)^{2} = \dfrac{36}{25}$

(iv) $\left(x^{\frac{4}{3}} y^{\frac{2}{3}}\right)^{-6} \left(x^{\frac{3}{5}} y^{\frac{4}{5}}\right)^{5} \underset{\text{rules of exps}}{=} x^{-\frac{24}{3}} y^{-\frac{12}{3}} x^{\frac{15}{5}} y^{\frac{20}{5}} \underset{\text{gather terms}}{=} x^{-8} x^{3} y^{-4} y^{4} = x^{-5}$

(v) $\dfrac{8x^{\frac{5}{2}} y^{\frac{4}{3}}}{4x^{\frac{7}{6}} y^{\frac{7}{4}}} \underset{\text{gather terms}}{=} \frac{8}{4} \cdot x^{\left(\frac{5}{2} - \frac{7}{6}\right)} y^{\left(\frac{4}{3} - \frac{7}{4}\right)} \underset{\text{simplify}}{=} 2x^{\frac{8}{6}} y^{-\frac{5}{12}} = 2x^{\frac{4}{3}} y^{-\frac{5}{12}}$

(vi) In this example we show how to rationalize the denominator of a more complicated expression by extending the logic used in (ii) above:

$$\frac{2}{1 - \sqrt{3}} = \frac{2}{1 - \sqrt{3}} \cdot \underbrace{\frac{1 + \sqrt{3}}{1 + \sqrt{3}}}_{=1}$$

Let's pause for a moment and ask why we chose to multiply by this particular expression. Recall that our objective is to get rid of the radicals on the denominator. In (ii) above we used the fact that multiplying a square root by itself eliminated the square root i.e.

$$\left(\sqrt{a}\right)^2 = a$$

Extending that logic, we need to multiply $1-\sqrt{3}$ by an expression which will 'annihilate' the square root. Our clue comes from the formula (see §3.5)

$$(a-b)(a+b) = a^2 - b^2 \tag{3.3.1}$$

Clearly, if $a = 1, b = \sqrt{3}$, then multiplying $1-\sqrt{3}$ by $1+\sqrt{3}$ will give $1^2-(\sqrt{3})^2 = -2$ and no more radicals! Of course, to prevent altering the original expression $\frac{2}{1-\sqrt{3}}$, we multiply *both* the numerator and the denominator by $1+\sqrt{3}$ so that in effect we multiply by **one** i.e. we change the way the expression *looks* - not it's value! It is important to understand the above logic - that way we can *extend* the ideas to e.g. cube roots (see (viii) below).

Back to the example!

$$\frac{2}{1-\sqrt{3}} = \frac{2}{1-\sqrt{3}} \cdot \underbrace{\frac{1+\sqrt{3}}{1+\sqrt{3}}}_{=1} = \frac{2(1+\sqrt{3})}{1-3} = -(1+\sqrt{3})$$

(vii) $\displaystyle \frac{\sqrt{3}}{\sqrt{6}-\sqrt{5}} \underset{\text{rat. denom.}}{=} \frac{\sqrt{3}}{\sqrt{6}-\sqrt{5}} \cdot \frac{\sqrt{6}+\sqrt{5}}{\sqrt{6}+\sqrt{5}} = \frac{\sqrt{3}(\sqrt{6}+\sqrt{5})}{6-5} = \sqrt{3}(\sqrt{6}+\sqrt{5})$

(viii) Consider the expression,

$$\frac{2}{\sqrt[3]{2}-\sqrt[3]{3}}$$

Suppose we are required to rationalize the denominator. Without the reasoning in (vi) above you might be tempted to multiply $\sqrt[3]{2}-\sqrt[3]{3}$ by $\sqrt[3]{2}+\sqrt[3]{3}$. This however, does not remove the radicals - in fact, it makes the expression more complicated! Instead, let us extend the *idea* that led to the use of (3.3.1) and consider the formula for the difference of two cubes:

$$(a-b)(a^2+ab+b^2) = (a^3-b^3)$$

OR (replacing a by $a^{\frac{1}{3}}$ and b by $b^{\frac{1}{3}}$)

$$(a^{\frac{1}{3}} - b^{\frac{1}{3}})(a^{\frac{2}{3}} + a^{\frac{1}{3}}b^{\frac{1}{3}} + b^{\frac{2}{3}}) = (a-b) \tag{3.3.2}$$

Letting $a = 2, b = 3$, we see that multiplying $\sqrt[3]{2}-\sqrt[3]{3}$ by $(2^{\frac{2}{3}} + 2^{\frac{1}{3}}3^{\frac{1}{3}} + 3^{\frac{2}{3}})$, by (3.3.2), leads to

$$(2^{\frac{1}{3}} - 3^{\frac{1}{3}})(2^{\frac{2}{3}} + 2^{\frac{1}{3}}3^{\frac{1}{3}} + 3^{\frac{2}{3}}) = (2-3) = -1 \tag{3.3.3}$$

Hence, we can eliminate the radicals in the denominator using (3.3.3). Following the procedure used above, we have,

$$\frac{2}{\sqrt[3]{2}-\sqrt[3]{3}}\cdot\underbrace{\frac{(2^{\frac{2}{3}}+2^{\frac{1}{3}}3^{\frac{1}{3}}+3^{\frac{2}{3}})}{(2^{\frac{2}{3}}+2^{\frac{1}{3}}3^{\frac{1}{3}}+3^{\frac{2}{3}})}}_{=1}=\frac{2(2^{\frac{2}{3}}+2^{\frac{1}{3}}3^{\frac{1}{3}}+3^{\frac{2}{3}})}{(2-3)}=-2(2^{\frac{2}{3}}+2^{\frac{1}{3}}3^{\frac{1}{3}}+3^{\frac{2}{3}})$$

It looks ugly, but we have successfully eliminated all radicals from the denominator!!! Notice how an understanding of the *reasons* behind a procedure makes it easier to generalise to 'new' situations.

In calculus, you will have to rationalize denominators and numerators containing variables rather than specific real numbers. The procedure is exactly the same!

Note 3.3.8

Expressions of the form b^x, b = constant > 0 are called *exponential functions.* They are defined $\forall x \in R$ - including irrational x e.g. $2^{\pi}, 2^{\sqrt{2}}$, etc. (see §7).

Exercises 3.3

1. Simplify

(a) $\dfrac{4ab^{-2}c^3}{a^{-3}b^3c^{-1}}$ (b) $\left(\dfrac{(2xz^{-2})^3(x^{-2}z)}{2xz^2}\right)^4$ (c) $(a^{-1}+b^{-2})^{-1}$

(d) $\left(\dfrac{3x^2y^{-2}}{2x^{-1}y^4}\right)^{-3}$ (e) $\dfrac{3(b^{-2}d)^4(2bd^3)^2}{(2b^2d^3)(b^{-1}d^2)^5}$ (f) $(x^{-1}-y^{-1})^{-1}$

2. Simplify

(a) $(x^2y^3)^{\frac{2}{5}}$ (b) $(3ab^2)^{\frac{2}{3}}$ (c) $\left(x^{\frac{1}{2}}+y^{\frac{1}{2}}\right)^{\frac{1}{2}}$ (d) $\dfrac{x^{\frac{1}{3}}y^{-\frac{3}{4}}}{x^{-\frac{2}{3}}y^{\frac{1}{2}}}$

(e) $\left(xy^{-\frac{2}{3}}\right)^3\left(x^{\frac{1}{2}}y\right)^2$ (f) $\left(\dfrac{a^{\frac{1}{2}}}{b^2}\right)^2\left(\dfrac{b^{\frac{3}{2}}}{a^{\frac{2}{3}}}\right)^3$

3. Rationalize the denominator

(a) $\dfrac{2}{\sqrt{3}-\sqrt{2}}$ (b) $\dfrac{1}{\sqrt{5}+\sqrt{3}}$

(c) $\dfrac{1}{\sqrt{x}-\sqrt{a}}$, a is a real positive constant.

3.4. POLYNOMIALS

A polynomial of *degree* $n = 0, 1, 2, 3, \ldots$, in the variable x, is given by an expression of the form

$$p(x) = a_n x^n + a_{n-1}x^{n-1} + \cdots + a_1 x + a_0$$

where, $a_0, a_1, \ldots, a_n \in R$ are the (constant) *coefficients.* For example,

$2x - 4$	degree one or linear polynomial
$x^2 - \frac{1}{2}x + 3$	degree two or quadratic polynomial
$\frac{1}{5}x^7 - 6x^3 + 2$	degree 7 polynomial
3	degree 0 polynomial

Note 3.4.1

Polynomials contain terms with only *positive integer* (including zero) powers of the variables.

Expressions such as

$$3x^{-\frac{1}{2}} + x^{\frac{1}{3}} + x^{-2} + 1$$

are not polynomials!

Addition Rule for Polynomials

Add (or subtract) 'like' terms i.e. those with the same power of the variable e.g.

$$\begin{aligned}
&(2x^4 + 6x^2 + x + 4) + (x^4 + \frac{1}{2}x^3 + x^2 + 3x) \\
= \; &(2+1)x^4 + (0 + \frac{1}{2})x^3 + (6+1)x^2 + (1+3)x + (4+0) \\
= \; &3x^4 + \frac{1}{2}x^3 + 7x^2 + 4x + 4
\end{aligned}$$

Similarly,

$$\begin{aligned}
&(2x^4 + 6x^2 + x + 4) - (x^4 + \frac{1}{2}x^3 + x^2 + 3x) \\
= \; &x^4 - \frac{1}{2}x^3 - 5x^2 - 2x + 4
\end{aligned}$$

Multiplication by a Constant

If

$$\begin{aligned}
p(x) &= x^4 - \frac{1}{2}x^3 - 5x^2 - 2x + 4 \\
3p(x) &= 3(x^4 - \frac{1}{2}x^3 - 5x^2 - 2x + 4) \\
&= 3x^4 - \frac{3}{2}x^3 - 15x^2 - 6x + 12
\end{aligned}$$

Multiplication of Polynomials

If a, b, c are polynomials:

$$\text{Distributive Property} \begin{cases} a(b+c) = & ab + ac \\ (a+b)c = & ac + bc \end{cases}$$

Example 3.4.2

(i) $2x(8 - 3x^2) = 16x - 6x^3$

(ii) $(x^2 - 3x + 2)(4x^3 + 3x - 1)$

$$= x^2 \left(4x^3 + 3x - 1\right) - 3x(4x^3 + 3x - 1) + 2\left(4x^3 + 3x - 1\right)$$
$$= 4x^5 + 3x^3 - x^2 - 12x^4 - 9x^2 + 3x + 8x^3 + 6x - 2$$
$$= 4x^5 - 12x^4 + 11x^3 - 10x^2 + 9x - 2$$

Multiplication Formulae

$$\begin{aligned} (x+y)^2 &= x^2 + 2xy + y^2 \\ (x-y)^2 &= x^2 - 2xy + y^2 \\ (x+y)(x-y) &= x^2 - y^2 \\ (x+y)^3 &= x^3 + 3x^2y + 3xy^2 + y^3 \\ (x-y)^3 &= x^3 - 3x^2y + 3xy^2 - y^3 \end{aligned} \tag{3.4.1}$$

The effective use of mathematical formulae is entirely dependent on interpretation. Consider, for example, the formula

$$(x+y)^2 = x^2 + 2xy + y^2 \tag{3.4.2}$$

and suppose we are required to expand the following expression

$$(\sqrt{x} + 2y)^2$$

To do this, think of x and y in (3.4.2) as *placeholders* (this is one of the most important concepts in mathematics) i.e. anything in the place of the x on the left-hand side of (3.4.2) is transferred to the corresponding place(s) occupied by that x on the right-hand side of (3.4.2). The same is true for the y. Hence, to perform the expansion of $(\sqrt{x} + 2y)^2$ we replace the $x - position$ by $\sqrt{x}$ and the $y - position$ by $2y$ in (3.4.1) i.e.

$$\begin{aligned} (\sqrt{x} + 2y)^2 &= \left(\sqrt{x}\right)^2 + 2\left(\sqrt{x}\right)(2y) + (2y)^2 \\ &= x + 4\sqrt{x}y + 4y^2 \end{aligned}$$

Try not to be confused by the x's and y's in the formula - read them only as *placeholders* i.e. (3.4.2) could equally be written as

$$(\theta + \Theta)^2 = \theta^2 + 2\theta\Theta + \Theta^2$$

or

$$(space1 + space2)^2 = (space1)^2 + 2(space1)(space2) + (space2)^2$$

The symbols are like *dummy variables* - only their positions are important.

All of the multiplication formulae in (3.4.1) (and, in fact, all mathematical formulae) should be interpreted in this way. You will see later that this will also facilitate the understanding of the theory of functions (see §4.3)

Note 3.4.3

If p and q are polynomials of degree m and n , respectively:

(i) The degree of the product pq is $m + n$.

(ii) The leading coefficient (the coefficient of the term with highest power) of pq is the product of the leading coefficients of p and q.

(iii) The constant term of pq is the product of the constant terms of p and q.

These observations are useful if say we wanted some quick information about the product of two polynomials without having to perform the actual multiplication (arises quite often in the theory of *limits* which is one of the early topics in beginning calculus). For example,

$$\begin{aligned} p(x) &= 17x^{10} + 3x^7 + \frac{2}{3}x^4 + x + 11 \\ q(x) &= 11x^{10} + 8x^5 + \frac{2}{9}x^3 + x^2 + \frac{3}{2} \end{aligned}$$

means that the polynomial $p(x)q(x)$ will have leading coefficient (coefficient of x^{10}) equal to $17 \cdot 11 = 187$, constant term $11 \cdot \frac{3}{2} = \frac{33}{2}$ and degree $10 + 10 = 20$.

Example 3.4.4

(i) $(3\sqrt{x} - 1)^2 = (3\sqrt{x})^2 - 2 \cdot 3\sqrt{x} \cdot 1 + 1^2 = 9x - 6\sqrt{x} + 1$, using (3.4.1)

(ii) $(x^{\frac{1}{3}} - x^{\frac{5}{2}})^3$

$$\begin{aligned} &= (x^{\frac{1}{3}})^3 - 3(x^{\frac{1}{3}})^2 x^{\frac{5}{2}} + 3x^{\frac{1}{3}}(x^{\frac{5}{2}})^2 - (x^{\frac{5}{2}})^3, \text{ using (3.4.1)} \\ &= x - 3x^{\frac{2}{3}}x^{\frac{5}{2}} + 3x^{\frac{1}{3}}x^5 - x^{\frac{15}{2}} \\ &= x - 3x^{\frac{19}{6}} + 3x^{\frac{16}{3}} - x^{\frac{15}{2}} \end{aligned}$$

Exercises 3.4

1. Perform the indicated operations and simplify.

   (a) $(3t-5)(4t-2)$
   (b) $1+y(y-2)$
   (c) $(3x+2)(4x^2+3x-1)$

   (d) $(x+10)^2$
   (e) $(x+8)(x-8)$
   (f) $(2t-5)^2$
   (g) $(2x^4+5x)(2x^4-5x)$
   (h) $[(t+2)+t^3]^2$
   (j) $[(t+2)+t^3][(t+2)-t^3]$
   (k) $(x+2)^3$
   (l) $(3u+1)^3$
   (m) $(u+4v)^3$

2. Simplify

   (a) $(\frac{1}{x}+\frac{1}{y})^2$ **(b)** $(x-y+z)^2$ **(c)** $(x+\frac{1}{x}+\frac{1}{x^2})(x-\frac{3}{x}+\frac{10}{x^2})$

3. Find the coefficient of x^3 in the expansion of $(x^2+2x+3)(x^3-3x^2+2x+1)$

4. If $a+b=1$, $a^2+b^2=2$, find a^3+b^3 (Hint: use the expansions of $(a+b)^3$ and $(a+b)^2$)

3.5. FACTORING

My old high-school mathematics teacher would often write an unfactored expression such as

$$x^2-x^3$$

on the blackboard and leave it there for the duration of the lesson. At the end of the class, he would ask how we 'felt' about this expression. Some of us admitted to having an irresistible urge to factor the polynomial into

$$x^2(1-x)$$

while others remarked that they had no particular opinion on the subject. My teacher would then point to the people in the former group (those with the urge to factor) and remark that they were "ready". It was all done in fun but just the same, it contained a very important message. My teacher was well aware that *fluency* in the basic techniques was essential for success in subsequent mathematics courses. This was his way of making that point. To him, an indication of the required *fluency* was when it didn't seem right to leave an expression in unfactored form - factoring had become second nature, almost automatic! To this day, the most successful students almost always exhibit this kind of *fluency*. It seems to be a necessary condition for success in mathematics. With this in mind, we review one of the most important techniques for simplification of mathematical expressions, particularly in relation to polynomials: *factoring.*

To *factor* a polynomial is to write it as the product of two or more polynomials of lesser degree e.g.

$$x^2 - 4 = \underbrace{(x+2)(x-2)}_{\text{FACTORS}}$$

$$x^3 - 6x^2 + 11x - 6 = (x-1)(x-2)(x-3)$$

Common Factors

This is the simplest factoring procedure - based on the distributive law (in reverse). For example, consider

$$4xy^2 - 16x^3y^4 + 8x^4y^2$$

Always factor out as much as you can. Proceed term-by-term i.e.

$4, -16, 8$ have common factor 2 but largest common factor 4.
x, x^3, x^4 have common factor x - which is also the largest common factor.
y^2, y^4, y^2 have common factor y but largest common factor y^2.

Hence,

$$4xy^2 - 16x^3y^4 + 8x^4y^2 = \underbrace{4xy^2(1 - 4x^2y^2 + 2x^3)}_{\text{No common factors - so finish!}}$$

Similarly,

$$4x^5y^2 - 12x^3y + 8x^2y^3 = 4x^2y(x^3y - 3x + 2y^2)$$

The multiplication formulae (3.4.1) can be used in reverse to provide some factoring formulae! This is illustrated in the following table.

$$
\begin{array}{rcl}
\text{To expand} & \rightarrow & \\
& \leftarrow & \text{To factor} \\
a(x+y+z) & = & ax+ay+az \\
(x+a)(x+b) & = & x^2+(a+b)x+ab \\
(x+y)^2 & = & x^2+2xy+y^2 \\
(x-y)^2 & = & x^2-2xy+y^2 \\
(x+y)(x-y) & = & x^2-y^2 \\
(x+y)^3 & = & x^3+3x^2y+3xy^2+y^3 \\
(x-y)^3 & = & x^3-3x^2y+3xy^2-y^3 \\
(x+y)(x^2-xy+y^2) & = & x^3+y^3 \\
(x-y)(x^2+xy+y^2) & = & x^3-y^3
\end{array} \tag{3.5.1}
$$

Note 3.5.1
In §3.4 we introduced the notion of a *placeholder* for effective interpretation of multiplication formulae. The formulae in (3.5.1), when interpreted in a similar fashion, can be used as simple factoring formulae as is illustrated in the next example.

Example 3.5.2
We will use the above formulae, (3.5.1) to factor certain 'non-standard' expressions. Note how the concept of *placeholder* continues to play a significant role.

(a) Consider the expression $4a^2 - 9b^2$. In order to proceed, we would like to write this in a form which 'fits' one of the formulae in (3.5.1). In fact, if we note that

$$4a^2 - 9b^2 = (2a)^2 - (3b)^2$$

we have a difference of two squares to which we can apply $(3.5.1)_5$ i.e.

$$4a^2 - 9b^2 = (2a)^2 - (3b)^2 = (2a+3b)(2a-3b)$$

(where we have used the fact that in the formula $(x+y)(x-y) = x^2 - y^2$,the '$x - position$' is taken by $2a$ and the '$y - position$' by $3b$ i.e. *placeholders.)*

(b) Similarly,

$$u^2 - 6u + 9 \underset{\text{trying to fit the form of } (3.5.1)_4}{=} u^2 - 2\cdot 3u + 3^2 \underset{\text{using } (3.5.1)_4}{=} (u-3)^2$$

(c) Similarly,

$$16x^2+24xy+9y^2 \underset{\text{fitting the form of } (3.5.1)_3}{=} (4x)^2+2\cdot 4x\cdot 3y+(3y)^2 \underset{\text{by } (3.5.1)_3}{=} (4x+3y)^2$$

(Here, the '$x - position$' in $(3.5.1)_3$ is occupied by $4x$ and the '$y - position$' by $3y$. Thinking in terms of *placeholders* avoids any possible confusion of the x's and y's in $(3.5.1)_3$ and in the expression to be factored)

Notes 3.5.3

(i) Answers can always be checked by multiplying out the factors!

(ii) Some polynomials cannot be factored if we limit ourselves to the set of real numbers. These polynomials are called *irreducible*.

Factoring Quadratics by Trial and Error

Consider the second degree polynomial (quadratic)

$$x^2 + cx + d \quad (c \text{ and } d \text{ are real constants})$$

If this quadratic can be factored (remember, it may be irreducible), we should be able to find real numbers a and b such that

$$x^2 + cx + d = (x + a)(x + b) = x^2 + (a + b)x + ab$$

i.e. we should be able to find $a, b \in R$ such that

$$\begin{aligned} a + b &= c \\ ab &= d \end{aligned} \tag{3.5.2}$$

(When we cannot solve (3.5.2), the quadratic is irreducible.) For example,

$$x^2 - 5x - 14 = (x + a)(x + b)$$

requires

$$ab = -14, \quad a + b = -5 \tag{3.5.3}$$

We have several possibilities for the first of these equations i.e.

$$\begin{aligned} a &= -14, \quad b = 1 \\ a &= 14, \quad b = -1 \\ a &= 7, \quad b = -2 \\ a &= -7, \quad b = 2 \qquad \surd \end{aligned}$$

However, only the last of these sums to -5, as required by (3.5.3). Hence,

$$x^2 - 5x - 14 = (x - 7)(x + 2)$$

Hence, to factor the quadratic $x^2 + cx + d$, we look for factors of d which sum to c. Once you have done a few, you'll be doing them in your head - remember, practice is the key!

Example 3.5.4

(a) $x^2 + x - 12$: Possibilities:

$$\begin{array}{r}(x-6)(x+2)\\(x+6)(x-2)\\(x-12)(x+1)\\(x+12)(x-1)\\(x-4)(x+3)\\(x+4)(x-3)\surd\end{array}$$

Again, we choose the combination that satisfies (3.5.2).

(b) $x^2 - 2x + 1$: Possibilities:

$$\begin{array}{r}(x+1)(x+1) = (x+1)^2\\(x-1)(x-1) = (x-1)^2\surd\end{array}$$

(c) $x^2 + x + 1$. There is no real number combination that satisfies (3.5.2) - so this quadratic is irreducible!

The more general quadratic

$$px^2 + cx + d, \qquad (p, c \text{ and } d \text{ are real constants})$$

can be dealt with in much the same way except that now we have also to account for factors of p i.e. we seek a decomposition of the form

$$px^2 + cx + d = (qx + a)(rx + b) = qrx^2 + (ar + bq)x + ab$$

where, this time, we require that the constants a, b, q and r satisfy,

$$\begin{aligned} ab &= d \\ qr &= p \\ ar + bq &= c \end{aligned} \qquad (3.5.4)$$

Instead of remembering this complicated set of equations, we apply the underlying logic i.e. the possibilities will consist of factors of p and factors of d which satisfy $(3.5.4)_3$. We can often get to the required factors quickly through trial and error.

Example 3.5.5

(a) $3x^2 - 5x - 2$: Some possibilities:

$$\begin{aligned}(3x-2)(x+1)\\(3x+2)(x-1)\\(3x-1)(x+2)\\(3x+1)(x-2)\surd\end{aligned}$$

The trick is to just keep trying all possibilities (there might be quite a few to go through!) until you get the required combination. As you can see from this example, we start with the factors of 3 and -2 and then choose the combination that satisfies $(3.5.4)_3$.

(b) $6x^2 - 7x + 2$: Some possibilities:

$$\begin{aligned}(6x-1)(x-2)\\(6x+1)(x+2)\\(6x+2)(x+1)\\(6x-2)(x-1)\\(3x+2)(2x+1)\\(3x-2)(2x-1)\surd\end{aligned}$$

(c) $2x^2 + x + 2$: This polynomial is irreducible - there are no real numbers q, a, r, b which satisfy (3.5.4).

We will return to our discussion of quadratic expressions in §3.7 where we will also discuss a quick and easy way to decide whether or not a quadratic is irreducible.

Factoring Higher Order Polynomials

Higher order polynomials are much more difficult to factor than quadratics (see §5.3) but there are some special cases which can be dealt with using formulae (3.5.1) - and the *placeholder* technique.

Example 3.5.6

(a) $8x^3 - 1 = (2x)^3 - 1^3 \underset{\text{using } (3.5.1)_9}{=} (2x-1)(4x^2+2x+1)$. This last quadratic is irreducible else we would continue by factoring the quadratic.

(b) $3x^4 + 10x^2 - 8$. If you look at this expression very carefully you will see it is basically a quadratic expression with the x replaced by an x^2 i.e. *a quadratic in*

x^2. Hence,

$$\begin{aligned} 3x^4 + 10x^2 - 8 = 3u^2 + 10u - 8, \text{ where } u = x^2 \\ = (3u-2)(u+4) \\ = (3x^2-2)(x^2+4) \text{ , since } u = x^2 \\ = (\sqrt{3}x - 2)(\sqrt{3}x+2)(x^2+4) \end{aligned}$$

Where, in the last line, we have used the formula $(3.5.1)_5$ for the difference of two squares and noted that the quadratic $x^2 + 4$ is irreducible.

(c) $x^6 - 64 =$

$$= (x^2)^3 - (2^2)^3 \underset{\text{using } (3.5.1)_9}{=} (x^2-4)\left[(x^2)^2 + (x^2)\cdot 2^2 + (2^2)^2\right]$$

$$= (x-2)(x+2)\left[u^2 + 4u + 16\right] \text{, where } u = x^2$$

The quadratic in u (or x^2) is irreducible, so we have, finally,

$$x^6 - 64 = (x-2)(x+2)\left(x^4 + 4x^2 + 16\right)$$

Factoring Expressions With Negative Exponents

This is one of the most important and frequently used techniques in beginning calculus - and one which continually causes difficulties! When factoring with negative exponents, there is one golden rule to remember:

Take out the lowest power!

Example 3.5.7

(a) $4x^{-3} - 6x^{-4} \underset{\text{lowest power is -4}}{=} 2x^{-4}(2x-3)$

(b) $x^{-\frac{5}{4}} + x^{\frac{3}{4}} \underset{\text{lowest power is } -\frac{5}{4}}{=} x^{-\frac{5}{4}}(1 + x^{\frac{8}{4}}) = x^{-\frac{5}{4}}(1+x^2)$

(c) $3x^2(x^2-4)^{\frac{3}{2}} - (x^2-4)^{\frac{5}{2}} =$ (there is a common factor of (x^2-4) and the lowest power is $\frac{3}{2}$)

$$\begin{aligned} &= (x^2-4)^{\frac{3}{2}}\left[3x^2 - (x^2-4)\right] \\ &= (x^2-4)^{\frac{3}{2}}\left[2x^2+4\right] \\ &= 2[(x-2)(x+2)]^{\frac{3}{2}}(x^2+2) \\ &= 2(x-2)^{\frac{3}{2}}(x+2)^{\frac{3}{2}}(x^2+2) \end{aligned}$$

Notice how we combine many techniques from the rules of exponents and factoring to arrive at the simplified expression.

(d) $-\frac{8}{9}x^2\,(x^2-1)^{-\frac{4}{3}} + \frac{4}{3}(x^2-1)^{-\frac{1}{3}} =$

(there is a common factor of (x^2-1) and the lowest power is $-\frac{4}{3}$)

$$\begin{aligned}
&= \frac{4}{9}(x^2-1)^{-\frac{4}{3}}[-2x^2+3(x^2-1)] \\
&= \frac{4}{9}(x^2-1)^{-\frac{4}{3}}(x^2-3) \\
&= \frac{4}{9}(x-1)^{-\frac{4}{3}}(x+1)^{-\frac{4}{3}}(x-\sqrt{3})(x+\sqrt{3})
\end{aligned}$$

(e) The following expression is part of an exercise from a beginning calculus course. The 'calculus part' has been applied and now the resulting expression must be simplified to obtain the required information:

$$\frac{(x+1)^{-2}(3x)(x^2-4)^{\frac{1}{2}} - (x+1)(-\frac{1}{2})(x^2-4)^{-\frac{1}{2}}}{(x^2-4)^2(x+1)}$$

$$= \frac{(x+1)^{-2}(\frac{1}{2})(x^2-4)^{-\frac{1}{2}}[6x(x^2-4)+(x+1)^3]}{(x^2-4)^2(x+1)}$$

Notice that in the numerator, the common factors are of the form $(x+1)$ and (x^2-4) with lowest powers -2 and $\frac{1}{2}$, respectively.

$$= \frac{[6x(x^2-4)+(x+1)^3]}{2(x^2-4)^{\frac{5}{2}}(x+1)^3}$$

$$= \frac{7x^3-21x+3x^2+1}{2(x^2-4)^{\frac{5}{2}}(x+1)^3}$$

(f) $x^{-6}+x^6+2=$

$$= x^{-6}\ \underbrace{[1+x^{12}+2x^6]}_{\text{quadratic in } x^6}$$

$$\begin{aligned}
&= x^{-6}[u^2+2u+1], \quad (\text{where } u = x^6) \\
&= x^{-6}(u+1)^2 \\
&= x^{-6}(x^6+1)^2
\end{aligned}$$

(g) $(x^2-1)^3(x+1)^2(x-1)+(x-1)(x+1)=$

$$\begin{aligned}
&= (x^2-1)^3(x+1)(x+1)(x-1)+(x^2-1) \\
&= (x^2-1)^4(x+1)\ +(x^2-1) \\
&= (x^2-1)[(x^2-1)^3(x+1)+1] \\
&= (x^2-1)x\left(x^6+x^5-3x^4-3x^3+3x^2+3x-1\right)
\end{aligned}$$

Notes 3.5.8

(i) The above examples are typical of expressions you will be required to factor in a beginning calculus course. You must know how to do this quickly and efficiently. To the calculus instructor, this type of mathematics is secondary and does not form part of the 'calculus material' - it is your responsibility to equip yourself with the necessary prerequisite techniques!

(ii) It is clear that the rules of exponents are extremely important when factoring expressions arising in calculus. At this point it is worth mentioning one particular part of the process which is often identified as troublesome. Consider the expression

$$(1-x)^{-\frac{5}{2}} + 3x(1-x)^{\frac{1}{2}}$$

The common factor is of the form $(1-x)$ and the lowest power is $-\dfrac{5}{2}$. Hence

$$(1-x)^{-\frac{5}{2}} + 3x(1-x)^{\frac{1}{2}} = (1-x)^{-\frac{5}{2}}\left[1+3x\cdot ?\right]$$

Students frequently encounter difficulties with what goes inside the bracket i.e. what takes the place of '?' This is like applying the rules of exponents in reverse. The resulting expression i.e.

$$(1-x)^{-\frac{5}{2}}\left[3x\cdot ?\right]$$

must give rise to the term with which we started i.e. $3x(1-x)^{\frac{1}{2}}$. So '?' must be of the form

$$(1-x)^{\text{something}}$$

That 'something' is what must be added to $-\dfrac{5}{2}$ to give $\dfrac{1}{2}$ i.e.

$$(1-x)^{\text{something } -\frac{5}{2}} = (1-x)^{\frac{1}{2}}$$

Hence the 'something' is $\dfrac{6}{2} = 3$. Thus,

$$(1-x)^{-\frac{5}{2}} + 3x(1-x)^{\frac{1}{2}} = (1-x)^{-\frac{5}{2}}\left[1+3x(1-x)^{3}\right]$$

If you are ever unsure, just multiply out the resulting expression and see if you get what you started with - that's the nice thing about factoring - you can always check because you know the answer!

(iii) Notice that much of the above factoring involves changing *sums to products.* Again, this is commonplace in calculus particularly when you want to find values of a variable e.g. x which makes a particular expression equal to zero (arises in curve sketching, maximum-minimum problems and in many other places).

(iv) Practice as much as you can with the following exercises - remember, in your calculus course you want to concentrate as much as possible on the (new) calculus theory - the precalculus techniques should be second nature!

Exercises 3.5

Factor and/or simplify

1. **(a)** x^2+5x **(b)** y^3+4y^2 **(c)** $8x^2y^3+16x^5y^4-24x^3y^6$
   (d) $(a-2b)(a+3b)(b-a)-(4a+5b)(a-b)(a-2b)$

2. **(a)** w^2-1 **(b)** x^2-4y^2 **(c)** $x^{100}-y^{50}$

3. **(a)** z^2-2z+1 **(b)** $9t^2+6t+1$ **(c)** $2w^2+32w+128$

4. **(a)** t^2-3t+2 **(b)** $z^2-3z-40$ **(c)** $x^2y^2-6xyz+9z^2$
   (d) $x^2-z^2-6xy+9y^2$ **(e)** $x^2+13x+42$ **(f)** $3x^2-12x+12$
   (g) $35x^2+29xy+6y^2$ **(h)** $6x^2-13xy-5y^2$ **(j)** x^3+8

5. **(a)** $(x+2)^3+(x-5)^3$ **(b)** $x^2+2xy+y^2+3x+3y+2$ (let $u=(x+y)$)

6. **(a)** $10x^3-5x^{-4}+15x^{-1}+20x$
   (b)
$$\frac{1}{2}\left(2x^2-x+1\right)^{-\frac{1}{2}}(4x-1)(x^3+1)^{\frac{1}{3}}+\left(2x^2-x+1\right)^{\frac{1}{2}}\frac{1}{3}(x^3+1)^{-\frac{2}{3}}(3x^2)$$
   (c)
$$\frac{3(x^2+1)^2(2x)(1-2x)^2-(x^2+1)^3(2)(1-2x)(-2)}{(1-2x)^4}$$

   Both parts 6(b) and 6(c) represent intermediate stages from solutions to questions posed in a beginning calculus assignment.

3.6. RATIONAL EXPRESSIONS

After polynomials, the next most common 'vehicle' for carrying mathematical information is a *rational expression* defined formally as a quotient of two polynomials i.e.

A *rational expression* is given by $\dfrac{p(x)}{q(x)}$ where p, q are polynomials $(q(x)\neq 0)$

For example,

$$\frac{x}{x+1},\quad \frac{3x^2+6}{x^2+2x},\quad \frac{x^2+3x}{6x^{15}+11} \tag{3.6.1}$$

Often, a rational expression may be in a form where relevant information is hidden. To see what we mean by this, consider an analogous expression from the set of real numbers i.e. a rational number $\frac{a}{b}$. This number has a description depending on the relative size of its numerator a and denominator b. In fact, we say that the rational number $\frac{a}{b}$ is proper when $a < b$ and improper otherwise e.g.

$$\frac{7}{2} \text{ is improper while } \frac{2}{7} \text{ is proper}$$

Proper expressions are simpler in the sense that they do not contain any 'hidden wholes' i.e. it is easier to judge the 'size' of a number if you write it in a form where the fractions are always proper e.g.

$$\frac{7}{2} = 3\frac{1}{2} \tag{3.6.2}$$

A similar classification can be made for rational expressions: when the **degree** of the numerator $p(x)$ is less than that of the denominator $q(x)$, we say that the rational expression is *proper*. Otherwise, we say it is *improper* e.g. in (3.6.1), the first two are improper rational expressions while the last one is proper. Furthermore, analogous to (3.6.2), we can always write an improper rational expression in terms of a proper one. For example,

$$\frac{x}{x+1} = 1 - \frac{1}{x+1} \tag{3.6.3}$$

(This decomposition is performed using synthetic or long division (see §5.2) and is an extremely important technique in beginning calculus). The decomposition (3.6.3) gives us quick information about the rational expression which might not be so obvious from its improper form e.g. as x becomes large, the rational expression tends towards the number 1. This is the type of information that you will need frequently in, for example, curve sketching.

Note 3.6.1

The golden rule regarding rational expressions is to avoid values of x which lead to division by zero (which, you may recall from §3.3, is something you must never do in the set of real numbers). Hence, in the rational expression

$$\frac{p(x)}{q(x)}$$

we always *exclude* x such that $q(x) = 0$ (We will revisit this when we consider the theory of functions in §4.3).

At this stage, it is important to become 'comfortable' with the algebra of rational functions i.e. become *fluent* in simplification, manipulation and interpretation. The rules are much the same as for the algebra of real numbers - but we must be careful to exclude values of x which lead to division by zero.

Example 3.6.2

(i) $\dfrac{x}{3x-6} - \dfrac{2}{2-x}$, $x \neq 2$

$$= \frac{x}{3(x-2)} - \frac{2}{2-x} = \frac{x}{3(x-2)} + \underbrace{\frac{2}{x-2}}_{=-\frac{2}{2-x}}$$

$$\underbrace{=}_{\text{taking common denominator}} \frac{x+6}{3(x-2)}$$

Notice how the value $x = 2$ has to be excluded to avoid division by zero.

(ii) $\dfrac{3x+6}{x^2+x-20}$

$$= \frac{3(x+2)}{(x+5)(x-4)} \text{ , } x \neq -5, 4 \qquad (3.6.4)$$

(iii) In the above two examples, it is clear from the form of the denominators in the simplified rational expressions that certain values of x would lead to division by zero and so should be avoided. e.g. in (3.6.4) it is clear that we should avoid the values $x = -5$ and $x = 4$ (it's as if there is a 'hole' there and it's clear that we should avoid it!). For this reason, it is often the case that the statement $x \neq -5, 4$ is simply omitted with the understanding that it is implied by the form of the denominator. In many cases however, in the process of simplifying a rational expression, factors are cancelled and it is no longer obvious from the final form of the expression which values if x should be avoided (the 'hole' is covered up!). In these cases, you should be careful to note at each stage, the 'troublesome' values of x and add the relevant information at the end. For example, consider the rational expression

$$\frac{x^2-1}{x+1}$$

It is clear that this expression is not valid (defined) at $x = -1$ since this value of x would result in division by zero. After noting that

$$\frac{x^2-1}{x+1} = \frac{(x-1)(x+1)}{x+1} = x - 1$$

one might be tempted to write

$$\frac{x^2-1}{x+1} = x - 1 \qquad (3.6.5)$$

Unfortunately, this is not strictly correct! (3.6.5) is true for all values of x *except* $x = -1$. In fact, the expression on the left-hand side of (3.6.5) (our original expression) is not even defined at $x = -1$ whereas the right-hand side of (3.6.5) is defined for all x. So the expressions in (3.6.5) are equivalent provided $x \neq -1$. A correct version of (3.6.5) would be

$$\frac{x^2 - 1}{x + 1} = x - 1 \, , \, x \neq -1$$

This is a very common mistake in calculus.

(iv) Consider now the expression

$$\frac{x^2 - 5x + 4}{2x + 6} \cdot \frac{x^2 + 5x + 6}{2x^2 - x - 1}$$

First note that each of the rational expressions in the product have values of x at which they are not defined (values of x which lead to division by zero). Factoring the quadratics in the expression leads to

$$\frac{(x-4)(x-1)}{2(x+3)} \cdot \frac{(x+3)(x+2)}{(2x+1)(x-1)}$$

It is clear that $x = -3, -\frac{1}{2}, 1$ have to be *excluded* immediately. We note this, bearing in mind that this might not be so obvious from the final simplified expression. In fact, if we cancel factors, we obtain

$$\frac{(x-4)(x+2)}{2(2x+1)}.$$

You will agree that it is not obvious from the simplified form of our expression that we should avoid the values $x = -3, 1$ - although it is clear that we should avoid $x = -\frac{1}{2}$. So to complete the simplification we write

$$\frac{(x-4)(x+2)}{2(2x+1)}, x \neq -3, -1 \tag{3.6.6}$$

(again, it is implied by the denominator that $x = -\frac{1}{2}$ should be avoided so there is no need to say so explicitly - although you can if you wish). You may also have noticed that if you substitute $x = -3$ or $x = -1$ in (3.6.6), you obtain well-defined numbers and not division by zero! This is true but remember that as in (iii) above, the expression in (3.6.6) is the same as our original expression at every value of x *except* -3 and -1. Hence, although you get 'sensible answers' from (3.6.6) at these values, they are not the values of the original expression at $x = -3$ and $x = -1$.

(v) If we exclude $x = 0$ and $x = 2$,

$$\frac{\frac{1}{x} - \frac{3}{2}}{\frac{2}{x-2} + \frac{5}{x}} = \frac{\frac{2-3x}{2x}}{\frac{2x+5(x-2)}{x(x-2)}}$$

$$= \frac{2-3x}{2x} \cdot \frac{x(x-2)}{7x-10}$$

$$= \frac{(2-3x)(x-2)}{2(7x-10)}\ , x \neq 0, 2$$

Example 3.6.3 - Expressions arising in Calculus

(i)

$$\frac{\frac{2}{x+h} - \frac{2}{x}}{h} = \frac{\frac{2x-2(x+h)}{x(x+h)}}{h}$$

$$= \frac{-2h}{hx(x+h)}$$

$$= \frac{-2}{x(x+h)}\ , h \neq 0$$

(ii) Here we modify the technique known as 'rationalization of the denominator' (Example 3.3.7) to rationalize the *numerator* of a rational expression:

$$\frac{\sqrt{x+h} - \sqrt{x}}{h}$$

$$= \frac{\sqrt{x+h} - \sqrt{x}}{h} \cdot \underbrace{\frac{\sqrt{x+h} + \sqrt{x}}{\sqrt{x+h} + \sqrt{x}}}_{=1}$$

$$= \frac{x+h-x}{h(\sqrt{x+h} + \sqrt{x})}$$

$$= \frac{h}{h(\sqrt{x+h} + \sqrt{x})}$$

$$= \frac{1}{(\sqrt{x+h} + \sqrt{x})}\ , h \neq 0$$

(as required i.e. no radical in the *numerator*)

Exercises 3.6

1. Simplify

(a) $\dfrac{x+6}{x^2-36}$ (b) $\dfrac{y^2+y}{5y+5}$ (c) $\dfrac{(x+2)^3}{x^2-4}$ (d) $\dfrac{5x}{x^2-4}+\dfrac{3}{x+2}$

2. Simplify

(a) $\dfrac{x+1}{x^2-4x+4}+\dfrac{4}{x^2+3x-10}$ (b) $\dfrac{x^2}{x^2-x+1}-\dfrac{x+1}{x}$

(c) $\dfrac{x^2-3x+2}{x^2-6x+8}$ (d) $\dfrac{\dfrac{x^2-1}{x^2-4}}{\dfrac{x^2-x-2}{x^2+x-2}}$

3. Simplify the following expressions

(a) $\dfrac{\dfrac{1}{(x+h)^2}-\dfrac{1}{x^2}}{h}$ (b) $\dfrac{\dfrac{1}{2x+2h+3}-\dfrac{1}{2x+3}}{h}$

4. Rationalize the denominator

(a) $\dfrac{3}{\sqrt{x}}$ (b) $\dfrac{3-2\sqrt{x}}{1+4\sqrt{x}}$

5. Rationalize the numerator

(a) $\dfrac{\sqrt{x+h+4}-\sqrt{x+4}}{h}$ (b) $\dfrac{\sqrt[3]{(x+h)}-\sqrt[3]{x}}{h}$

(c) $\dfrac{\dfrac{1}{\sqrt{(x+h)^2}}-x^{-1}}{h}$, $x+h>0$.

3.7. QUADRATIC EQUATIONS

The ability to factor and simplify polynomial and rational expressions is particularly important when we need to solve *equations* which contain these types of expressions. In this section we review techniques for solving the simplest type of polynomial equation - the quadratic equation.

Recall that a *linear* (or first degree) equation

$$ax+b=0 \ , \ a,b \in R$$

has exactly one solution $x = -\frac{b}{a}$, $a \neq 0$ (this represents where the graph of the line $y = ax + b$ intersects the $x - axis$. A *quadratic* (or second degree) equation takes the form

$$ax^2 + bx + c = 0, \quad a, b, c \in R, \quad a \neq 0 \qquad (3.7.1)$$

Some examples of quadratic equations are

$$\begin{aligned} x^2 - 4 &= 0 \text{ with solutions } x = \pm 2 \\ (x-1)^2 &= 0 \text{ with solution(s) } x = 1 \text{(twice)} \\ x^2 + 3x + 2 &= 0 \text{ with solutions } x = -1, -2 \\ x^2 + x + 1 &= 0 \text{ with no (real) solutions} \end{aligned}$$

Solutions of (3.7.1) are often referred to as *roots* or *zeros* of the quadratic expression (mainly because they represent points where the graph of the quadratic intersects the $x - axis$ - see §5.1). The process by which we arrive at solutions of (3.7.1) is not as straightforward as in the case of the linear equation. In fact, it is clear from the above examples that there are a number of possibilities - including the case where the quadratic equation has no real solutions! What we need is a test to tell us if (3.7.1) has real solutions and if so, what form they take i.e. are they distinct or coincident? As a first attempt, consider the equation

$$x^2 - 4 = 0 \Longleftrightarrow (x-2)(x+2) = 0 \Longleftrightarrow x = \pm 2$$

This is based on the simple rule

$$uv = 0 \Longleftrightarrow u = 0 \text{ or } v = 0$$

i.e. if you can factor an expression the solution of the equation comes from the two (simpler) equations which arise when each of the factors is equated to zero e.g.

$$\begin{aligned} x^2 - x - 6 &= 0 \Longleftrightarrow (x-3)(x+2) = 0 \Longleftrightarrow x - 3 = 0 \text{ or } x + 2 = 0 \Longleftrightarrow x = -2, 3 \\ x^2 - 2x + 1 &= 0 \Longleftrightarrow (x-1)^2 = 0 \Longleftrightarrow (x-1) = 0 \text{ or } (x-1) = 0 \Longleftrightarrow x = 1 \text{(twice)} \end{aligned}$$

Hence, if we can factor the quadratic expression, the solutions will follow from the factors.

Example 3.7.1
Solve

(i) $6x^2 - 5x = -1$ **(ii)** $3x^2 = x - 1$

Solution

(i) $6x^2 - 5x + 1 = 0 \Longleftrightarrow (3x-1)(2x-1) = 0 \Longleftrightarrow x = \frac{1}{3}, \frac{1}{2}$

(ii) $3x^2 - x + 1 = 0$, - can't seem to find any factors by inspection!

Part (ii) of Example 3.7.1 illustrates the main problem with the 'factoring by inspection' method of solving quadratic equations: what do we do if we cannot factor the quadratic? Clearly, we need a better method for deciding whether or not (3.7.1) has real solutions and, if so, how to find them.

Recall that all solutions of (3.7.1)(including those obtained through the 'factoring by inspection' method) can be expressed in the form

$$x = \frac{-b \pm \sqrt{b^2 - 4ac}}{2a} \tag{3.7.2}$$

(We will discuss the proof of this formula later). It is clear that there are three possible cases arising from (3.7.2).

CASE 1 : $b^2 - 4ac > 0$. In this case, (3.7.2) gives two,real, distinct solutions of (3.7.1)

CASE 2 : $b^2 - 4ac = 0$. In this case, (3.7.2) gives two, real coincident solutions of (3.7.1)

CASE 3 : $b^2 - 4ac < 0$. In this case, (3.7.2) gives two, non-real, distinct solutions of (3.7.1)

In CASE 3, (3.7.2) requires us to take the square root of a negative number. Hence, the solutions cannot be real but they can be interpreted as complex numbers (more on this later). Consequently, when solving quadratic equations we proceed as follows:

1. Put the equation in the form (3.7.1)
2. Check $b^2 - 4ac$ to see which of the three cases applies.
3. If there are real solutions, find them either by factoring the left-hand side of (3.7.1) by inspection (if possible) or by using (3.7.2) directly.

Note 3.7.2

CASE 1 corresponds to the case where the graph of the quadratic expression intersects the $x - axis$ in two distinct places.

CASE 2 corresponds to the case where the graph of the quadratic expression touches (is tangent to) the $x - axis$ at one particular place.

CASE 3 corresponds to the case where the graph of the quadratic expression never intersects the $x - axis$ i.e. the quadratic expression is never zero and so must be either always positive or always negative.

Example 3.7.3

(i) $5x^2 - 3x - 6 = 0$. Comparing with (3.7.1),

$$b^2 - 4ac = (-3)^2 - 4(5)(-6) = 129 > 0$$

hence we expect two, real distinct solutions. In fact, they are

$$x_1 = \frac{3 + \sqrt{129}}{10}, \quad x_2 = \frac{3 - \sqrt{129}}{10}$$

(ii)

$$x^2 + 3x + 2 = -5 \tag{3.7.3}$$

First put this in the form (3.7.1).

$$x^2 + 3x + 7 = 0$$

Comparing with (3.7.1), $b^2 - 4ac = (3)^2 - 4(1)(7) = -19 < 0$, hence there are no real solutions of (3.7.3)!

Warning!

Always put the quadratic equation in (standard) form (3.7.1) (i.e. with a zero on the right-hand side) before applying (3.7.2) e.g. had we not put (3.7.3) in the form (3.7.1) before applying (3.7.2), we would have come to the wrong conclusion i.e. applying (3.7.2) to the left-hand side of (3.7.3) leads to

$$x = \frac{-3 \pm \sqrt{1}}{2}$$

i.e. two real solutions (which are in fact the solutions of the equation

$$x^2 + 3x + 2 = 0$$

and not the solutions of (3.7.3)) and the **wrong** conclusion!

(iii) $2x^2 + x + 3 = 4$. In the form (3.7.1), this becomes

$$2x^2 + x - 1 = 0 \tag{3.7.4}$$

Hence, $b^2 - 4ac = 9 > 0$, hence, we expect two, real distinct solutions. In fact, from (3.7.2), we obtain

$$x_1 = -1, \quad x_2 = \frac{1}{2} \tag{3.7.5}$$

This example illustrates another important point worth highlighting at this stage. You may have noticed that the left-hand side of (3.7.4) can be factored by inspection so that the solutions (3.7.5) could have been obtained directly without using (3.7.2) i.e. as we mentioned above, factoring the quadratic *expression* leads to the solutions of the quadratic *equation.* The converse is also true! i.e. suppose we were asked to factor the quadratic that appears on the left-hand side of (3.7.4) i.e.

$$2x^2 + x - 1$$

and we could not 'see' the factors by inspection. Solving the equation (3.7.4) leads to the roots (3.7.5) which must therefore correspond to the equations (using the 'factoring by inspection' method *backwards)*

$$(x + 1) = 0 \text{ and } (x - \frac{1}{2}) = 0$$

Hence, (3.7.4) must take the form

$$(x + 1)(x - \frac{1}{2}) = 0 \quad \underset{\text{multiply both sides by 2}}{\Longleftrightarrow} \quad (2x - 1)(x + 1) = 0$$

In other words,

$$2x^2 + x - 1 = (2x - 1)(x + 1)$$

and we have factored the quadratic! The point is that factoring quadratics and solving the corresponding quadratic equation are basically equivalent i.e. *you can use one to do the other.* Hence,

CASE 1: $ax^2 + bx + c = (x - x_1)(x - x_2)$ where x_1, x_2 are the distinct, real solutions of (3.7.2).
CASE 2: $ax^2 + bx + c = (x - x_1)^2$ where $x_1$is the repeated, real solution of (3.7.2).
CASE 3: $ax^2 + bx + c$ is called *irreducible* i.e. no real factors - since there are no real solutions of (3.7.2).

Example 3.7.4
Factor the quadratic

$$x^2 - 3x + 1$$

Solution
No obvious factors so let us solve the corresponding quadratic equation

$$x^2 - 3x + 1 = 0$$

From (3.7.2),

$$x_1 = \frac{1}{2}(3 + \sqrt{5}), \quad x_2 = \frac{1}{2}(3 - \sqrt{5}) \quad \text{(two real, distinct roots)}$$

Thus,

$$x^2 - 3x + 1 = (x - x_1)(x - x_2)$$

gives the required factors!

Note 3.7.5

The proof of (3.7.2) is based on a very important technique known as *completing the square.* Consider the equation

$$x^2 - 6x = -2 \tag{3.7.6}$$

Write the left-hand side of (3.7.6) in terms of a perfect square i.e.

$$x^2 - 6x = (x - 3)^2 - 9$$

(you can verify this by expansion - in fact,

$$x^2 + px = (x + \frac{p}{2})^2 - (\frac{p}{2})^2, \text{ for } p \in R \text{)} \tag{3.7.7}$$

Hence, (3.7.6) becomes

$$(x - 3)^2 = 7$$

So that

$$x - 3 = \pm\sqrt{7}$$

and

$$x = 3 \pm \sqrt{7}$$

which could have been obtained using the quadratic formula (3.7.2). In fact, the method of completing the square (i.e. using (3.7.7)) works on any quadratic equation but the quadratic formula (3.7.2) uses the technique once and for all on the general quadratic equation (3.7.1) (thereby removing the need to do it for each specific case) leading to a nice simple formula for the solutions (roots).

Quadratics with Complex (Nonreal) Roots

We have already noted that in CASE 3 i.e. when

$$b^2 - 4ac < 0$$

the corresponding quadratic equation (3.7.1) has no *real* solutions. Consequently the graph of the quadratic never intersects the $x - axis$ i.e. the quadratic cannot change sign and so is either always positive or always negative. In fact, in this case, the quadratic equation (3.7.1) has *complex (nonreal) solutions* and, as we shall see below, they always occur in *conjugate pairs* i.e. it is not possible for a quadratic expression to have one real and one complex (nonreal) root!

Before we actually write down a representation of the complex solutions of (3.7.1), let us undertake a brief review of *complex numbers.*

The complex number system was devised to deal with e.g. the square root of a negative number i.e. if $a > 0$,

$$\begin{aligned}\sqrt{-a} &= \sqrt{i^2 a} \text{ (writing } i^2 = -1) \\ &= i\sqrt{a}\end{aligned}$$

which is known as the *principal square root* of $-a$. e.g.

$$\sqrt{-4} = \sqrt{4i^2} = \sqrt{4} \cdot \sqrt{i^2} = 2i$$

the principal square root of -4. Similarly, $\sqrt{-16} = 4i$, $\sqrt{-\frac{1}{9}} = \frac{i}{3}$, etc.

In general, the set of all complex numbers, denoted by C consists of numbers z of the form

$$z = a + ib, \quad a, b \in R, \quad i^2 = -1$$

The real number a is referred to as the *real part* of the complex number z and denoted by $\text{Re}(z)$. The real number b is referred to as the *imaginary part* of the complex number z and is denoted by $\text{Im}(z)$.

Notes 3.7.6

(i) $R \subset C$ i.e. all real numbers are complex - in the sense that real numbers are basically complex numbers with zero imaginary part (i.e. $b = 0$, $\forall z \in R$).

(ii) The complex number $\bar{z} = a - ib$ is called the *complex conjugate* of $z = a + ib$.

Returning now to the quadratic equation (3.7.1) with $b^2 - 4ac < 0$, we can write for its complex (nonreal) solutions,

$$\begin{aligned}x &= \frac{-b \pm \sqrt{b^2 - 4ac}}{2a} \\ &= \frac{-b \pm i\sqrt{4ac - b^2}}{2a}\end{aligned} \tag{3.7.8}$$

Notice that, as stated above, the complex (nonreal) solutions always occur in conjugate pairs.

Example 3.7.7

$$x^2 + 2x + 2 = 0 \Longleftrightarrow x = \frac{-2 \pm \sqrt{-4}}{2} = \frac{-2 \pm \sqrt{4i^2}}{2} = \frac{-2 \pm 2i}{2} = -1 \pm i$$

Warning!

We cannot interpret these nonreal solutions as intercepts on an axis in the Cartesian plane as we did for the case of real solutions (we have already established above that CASE 3 quadratic expressions have graphs which never intersect the $x-axis$ in the Cartesian plane). However, using the analogy for real factors, we can write the quadratic as a product of complex (nonreal) factors i.e.

$$x^2 + 2x + 2 = (x - (-1 + i))(x - (-1 - i))$$

Exercises 3.7

1. Solve by factoring

   (a) $x^2 = 3x$ **(b)** $x^2 - \frac{9}{4} = 0$ **(c)** $x^2 - 3x - 10 = 0$

   (d) $x^2 + 13x = -22$

2. Factor using the formula (3.7.2)

   (a) $x^2 + 8x + 12$ **(b)** $x^2 + 5x + 3$ **(c)** $\frac{1}{4}x^2 + \frac{1}{3}x - \frac{1}{6}$

   (d) $10^{-4}w^2 + 2 \times 10^{-3}w + 10^{-2}$

3. Solve using any method

   (a) $x^2 - 3x = 4x - 10$ **(b)** $z^2 - 3z + 4 = 2z^2 + 5z - 7$

   (c) $w(w + 2) = (w - 1)^2$

4. Write in the form bi, $b \in R$.

   (a) $\sqrt{-80}$

   (b) $\sqrt{-100}$

5. Write the conjugate

   (a) $10 + i\sqrt{53}$

   (b) $\dfrac{1 + i\sqrt{3}}{4}$

6. Find all solutions

   (a) $x^2 + 4 = 0$

   (b) $3x^2 + 2x + 5 = 0$

   (c) $8v^2 + 4v + 3 = 0$

3.8. INEQUALITIES, INTERVALS AND THE TEST-POINT METHOD

In calculus, it is often the case that one has to determine values of a variable x, say, which satisfy a certain *in*equation (or inequality) e.g. values of x for which the graph of the function $f(x)$ is: (i) positive (ii) increasing (iii) decreasing (iv) concave up etc. In this section, we discuss methods for the solution of certain inequalities arising in calculus. In doing so, we note the importance of factoring techniques and techniques for solving the related quadratic equation. First we review some basic notation:

Intervals
Let $a, b \in R$.

$$\begin{aligned}
(a,b) &= \{x : a < x < b\}\ ; \qquad [a,b] = \{x : a \le x \le b\} \\
[a,b) &= \{x : a \le x < b\}; \qquad (a,b] = \{x : a < x \le b\} \\
(a,\infty) &= \{x : x > a\}; \qquad [a,\infty) = \{x : x \ge a\} \\
(-\infty,b) &= \{x : x < b\}; \qquad (-\infty,b] = \{x : x \le b\} \\
(-\infty,\infty) &= R, \text{ the set of real numbers}
\end{aligned}$$

Notes 3.8.1

(i) $(\cdot,\cdot)$ denotes an open interval e.g. the set (a,b) does not contain its end-points a and b.

$[\cdot,\cdot]$ denotes a closed interval e.g. the set $[a,b]$ does contain its end-points a and b.

$(\cdot,\cdot]$ denotes an interval which is open at the left-hand end and closed at the right e.g. $(a,b]$ contains the right-hand end-point b but not the left-hand end-point a.

$[\cdot,\cdot)$ denotes an interval which is open at the right-hand end and closed at the left e.g. $[a,b)$ contains the left–hand end-point a but not the right-hand end-point b.

(ii) $\pm\infty$ are not numbers! - merely notation to denote $\pm$ infinity i.e. very large in either direction. For that reason we should *never* close $\pm\infty$ - they do not exist as numbers so they cannot be included in any set i.e. it is **inappropriate** to write e.g.

$$[-\infty, b) \qquad \text{or} \qquad [a, \infty]$$

Linear Inequalities

Just as equations are composed of mathematical expressions linked by an 'equality'or 'equals sign', inequations or inequalities are similarly mathematical expressions linked by inequalities or signs such as $<, >, \leq$ or $\geq$. A *linear inequality* is one in which the mathematical expressions appearing are linear e.g.

$$3x + 2 < 6x - 5$$

Essentially, inequalities are handled in much the same way as equations. There is one major difference however which we note among the following 'ground rules' below. Let $a, b, c \in R$:

1. If $a < b$ and $b < c$ then $a < c$.
2. If $a < b$ then $a + c < b + c$.
3. If $a < b$ and $c < 0$ then $ac > bc$!!!!!!!!!! * * * * * * * *
4. If $a < b$ and $c > 0$ then $ac < bc$

The third property is the one to note - it is peculiar to inequalities. In words,

> Whenever you multiply both sides of an inequality by a *negative* number, you must change the direction of the inequality!

Example 3.8.2

(i) $-2x + 6 < 18 + 4x \iff -6x < 12 \iff -x < 2 \underset{\textbf{careful!}}{\iff} x > -2$ or $x \in (-2, \infty)$.

(ii) $-3 < \dfrac{7 - 2x}{3} \leq 4$. This is a 'double-sided' inequality so we consider both sides separately.

$$\begin{aligned}
&\iff -9 < 7 - 2x \leq 12. \\
&\iff -9 < 7 - 2x \quad \text{and} \quad 7 - 2x \leq 12 \\
&\iff -16 < -2x \quad \text{and} \quad -2x \leq 5 \\
&\iff -8 < -x \quad \text{and} \quad -x \leq \frac{5}{2} \\
&\iff 8 > x \quad \text{and} \quad x \geq -\frac{5}{2} \\
&\iff x < 8 \quad \text{and} \quad x \geq -\frac{5}{2} \\
&\iff x \in [-\frac{5}{2}, 8)
\end{aligned}$$

Quadratic Inequalities and the Test-Point Method

Quadratic inequalities take the form:

$$\begin{aligned} ax^2+bx+c &> 0 \\ &\geq 0 \\ &< 0 \\ &\leq 0, \qquad a,b,c \in R, \quad a \neq 0 \end{aligned} \tag{3.8.1}$$

We shall see that the solution of these inequalities relies heavily on results obtained from the theory of quadratic equations. In fact, we know from §3.7 that the quadratic equation

$$ax^2+bx+c=0$$

has either two distinct real solutions, two coincident real solutions or no real solutions. This gives three possibilities for solving quadratic inequalities:

I *Quadratic Expression has two distinct real roots* ($b^2-4ac>0$ in (3.8.1))

This is the most interesting case. We illustrate by example and, in doing so, we introduce the 'Test-Point Method' for solving inequalities of this type. Consider the quadratic inequality

$$x^2-2x-3>0$$

The first thing to do is to determine whether the left-hand side can be factored. In fact, $b^2-4ac=16>0$ and we expect two real and distinct roots. We obtain

$$x^2-2x-3=(x+1)(x-3)>0 \tag{3.8.2}$$

We recall (from §3.7) that the roots $x=-1,3$ represent where the graph of the quadratic intersects the $x-axis$. This means that the quadratic can change sign (i.e. pass above or below the $x-axis$) **only** through these values of x. For this reason, we refer to $x=-1,3$ as the *dividing points*. This means that the quadratic expression cannot change sign in any of the regions

$$x<-1, \quad -1<x<3, \qquad x>3 \tag{3.8.3}$$

For example, if the expression is positive for *any* value of x in $x<-1$, it must stay positive for *all* values of x in this interval (same for the intervals $(-1,3)$ and $(3,\infty)$. Hence, we can easily determine the sign of the quadratic expression in each of the three intervals in (3.8.3) by simply choosing a 'test-point' (any *arbitrary* point) in each interval and noting the sign of the quadratic expression at that point (and hence at every point in that interval!). When we've done this for all three intervals, we simply 'pick-off' the intervals which have the desired sign i.e. that required by the inequality, in this case (3.8.2).

We summarize using the following very simple diagram:

-1		3	

$x < -1$	$-1 < x < 3$	$x > 3$
t.p. $x = -2$	t.p. $x = 0$	t.p. $x = 4$
$(x+1)(x-3)$	$(x+1)(x-3)$	$(x+1)(x-3)$
$- \cdot - = +$	$+ \cdot - = -$	$+ \cdot + = +$
> 0	< 0	> 0

Hence,

$$(x+1)(x-3) > 0 \text{ when } x \in (-\infty, -1) \cup (3, \infty)$$

and, incidentally

$$(x+1)(x-3) < 0 \text{ when } x \in (-1, 3)$$

i.e. we get this additional result without any 'extra work'.

Notes 3.8.3

(i) Had we considered the inequality

$$x^2 - 2x - 3 \geq 0$$

we would have included the roots of the quadratic in the solution set i.e.

$$x \in (-\infty, -1] \cup [3, \infty)$$

(ii) For a quadratic expression such as that appearing in (3.8.2) (with two, distinct, real roots) we expect the sign to alternate as the quadratic passes through the dividing points (this is evident from the following graph of the curve).

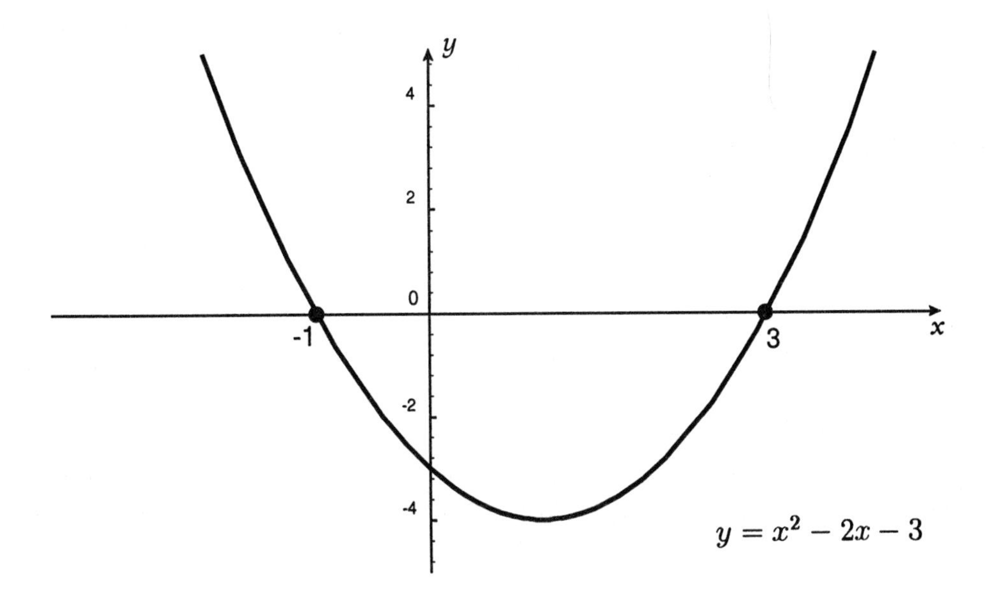

- hence, we really need only check the sign in the first interval and alternate the sign in subsequent intervals. However, when dealing with more complicated expressions e.g. rational expressions(see later) one should not make such an assumption and ensure that the sign is checked in each and every interval.

(iii) The Test-Point Method will prove to be most useful when we consider inequalities involving higher order polynomials and rational functions (later).

Example 3.8.4
Consider the inequality

$$\begin{aligned} 2x^2 &\leq 1-x \\ &\Longleftrightarrow 2x^2+x-1 \leq 0 \underset{\text{Factor by inspection}}{\Longleftrightarrow} (2x-1)(x+1) \leq 0 \end{aligned}$$

The dividing points are $x=-1$ and $x=\dfrac{1}{2}$

$x<-1$	$-1<x<\frac{1}{2}$	$x>\frac{1}{2}$
t.p. $x=-2$	t.p. $x=0$	t.p. $x=4$
$(2x-1)(x+1)$	$(2x-1)(x+1)$	$(2x-1)(x+1)$
$-\cdot-=+$	$-\cdot+=-$	$+\cdot+=+$
>0	<0	>0

(dividing points on the line: -1, $\frac{1}{2}$)

Hence,

$$(2x-1)(x+1) \leq 0 \Longleftrightarrow x \in [-1, \frac{1}{2}]$$

(and

$$(2x-1)(x+1) > 0 \Longleftrightarrow x \in (-\infty,-1) \cup (\frac{1}{2},\infty) \quad)$$

II *Quadratic Expression has two coincident real roots* ($b^2-4ac=0$ in (3.8.1))

Let the root of the quadratic be $x=a$. Since the root is repeated, we can always rewrite the quadratic as the perfect square $(x-a)^2$ - leading to four possible cases

$$\begin{aligned} (x-a)^2 &> 0, && \text{true } \forall x \text{ except } x=a. \\ &\geq 0, && \text{true } \forall x. \\ &< 0, && \text{never true (a square cannot be negative!)} \\ &\leq 0, && \text{true only for } x=a. \end{aligned} \tag{3.8.4}$$

Example 3.8.5
Consider the inequality:

$$
\begin{aligned}
6x &< x^2 + 9 \\
&\iff x^2 - 6x + 9 > 0 \quad (b^2 - 4ac = 0 \text{ i.e. quadratic has 2 real coincident roots}) \\
&\iff (x-3)^2 > 0
\end{aligned}
$$

which is true $\forall x$ except $x = 3$ by (3.8.4).

III *Quadratic Expression Has No Real Roots* ($b^2 - 4ac < 0$ in (3.8.1))

In this case, the quadratic expression cannot change sign - it has no real roots so its graph never intersects the $x - axis$ and so must remain either above or below the $x - axis$ *for all values* of x.

Example 3.8.6
Consider the inequality

$$x^2 + 2x + 2 > 0$$

Noting that $b^2 - 4ac = -4 < 0$ for the quadratic $x^2 + 2x + 2$, we conclude that it has no real roots. Hence, by what has been said above, $x^2 + 2x + 2$ cannot change sign. To find out which sign it has, choose any value of x e.g. $x = 0$. Clearly,

$$\text{at } x = 0, \quad x^2 + 2x + 2 = 2 > 0$$

Hence,

$$x^2 + 2x + 2 > 0, \quad \forall x \in R$$

This is clear from the graph of the quadratic:

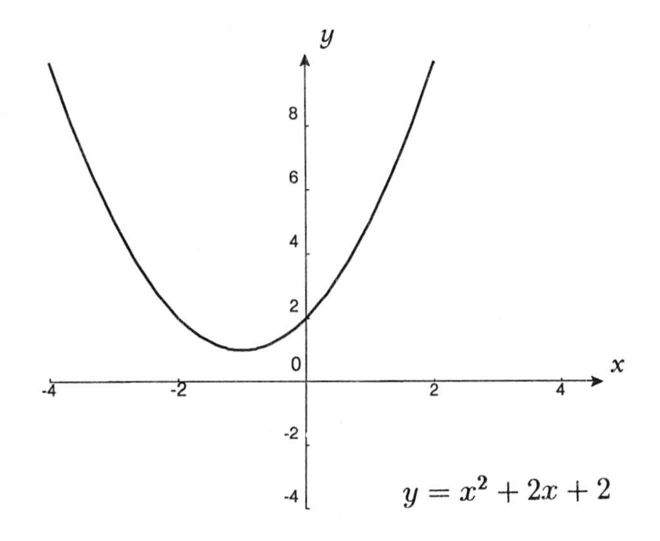

In general, when the quadratic expression in (3.8.1) has no real roots we can say the following:

$$\begin{aligned} \text{When } c &> 0, \quad ax^2+bx+c>0, \quad \forall x \in R \\ \text{When } c &< 0, \quad ax^2+bx+c<0, \quad \forall x \in R \end{aligned}$$

(Remember, $ax^2 + bx + c \neq 0$, for any value of $x \in R$.)

Other Inequalities

The test-point method works on any polynomial inequality, provided the corresponding polynomial expression can be factored (we will discuss the factoring of higher order polynomials later - in §5.3).

Example 3.8.7

Consider the (cubic) inequality

$$(x+2)(x-1)(x-4) < 0$$

Clearly the dividing points are

$$x = -2, 1, 4$$

Applying the test-point method we obtain,

	-2		1		4	
$x < -2$		$-2 < x < 1$		$1 < x < 4$		$x > 4$
t.p. $x = -3$		t.p. $x = 0$		t.p. $x = 3$		t.p. $x = 5$
cubic		*cubic*		*cubic*		*cubic*
$- \cdot - \cdot - = -$		$+ \cdot - \cdot - = +$		$+ \cdot + \cdot - = -$		$+ \cdot + \cdot + = +$
< 0		> 0		< 0		> 0

Hence,

$$\begin{aligned} (x+2)(x-1)(x-4) &< 0 \iff x \in (-\infty, -2) \cup (1, 4) \\ (\text{and } (x+2)(x-1)(x-4) &\geq 0 \iff x \in [-2, 1] \cup [4, \infty)) \end{aligned}$$

Rational inequalities (inequalities involving rational expressions) are also common in beginning calculus. The test-point method is easily adapted to deal with these inequalities - but we have to be careful! Consider the following example.

Example 3.8.8

Consider the inequality

$$\frac{3}{x} > 5$$

It is tempting to multiply through by x to 'clear the fraction' i.e.

$$3 > 5x \tag{3.8.5}$$

and conclude that

$$x < \frac{3}{5}$$

Unfortunately, this is not the 'whole story'. To obtain (3.8.5), we have assumed that $x > 0$ (remember, if $x < 0$, we have to reverse the direction of the inequality!). What about the case $x < 0$? (we have already ruled out $x = 0$ because of the requirement that we cannot divide by zero - see §3.6). In fact, we can cater for all cases using a very simple procedure which makes use of the test-point method (it is worth noting at this stage that cross-multiplication in inequalities involving rational expressions should be avoided at all costs!):

STEP 1: Make a zero on the right-hand side

$$\frac{3}{x} - 5 > 0$$

STEP 2: Combine terms

$$\frac{3 - 5x}{x} > 0$$

STEP 3: Identify the dividing points.

To do this we again look at the 'roots' of the expression on the left-hand side of the inequality but this time we also include the values of x where that same expression becomes undefined i.e. where division by zero occurs (a rational expression can change sign not only where it intersects the $x - axis$, but also where it becomes undefined i.e. at a *vertical asymptote* - you will encounter this term again in your beginning calculus course). To see this consider the following graph of the above rational expression.

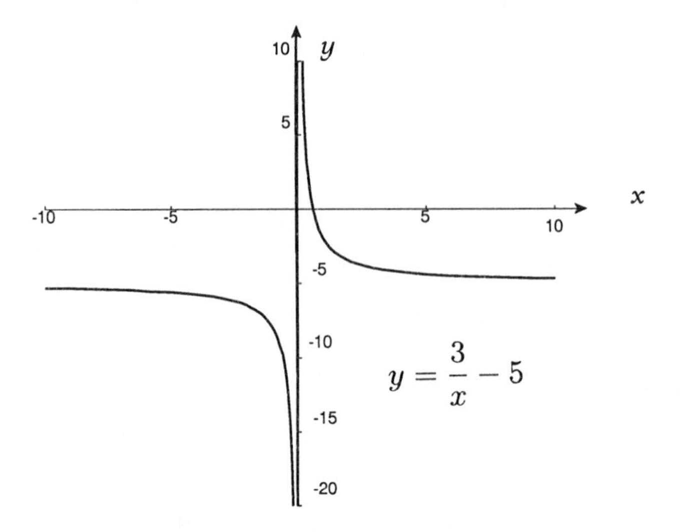

Hence we identify $x = 0, \frac{3}{5}$ as our dividing points. Applying the (extended) test-point method in the corresponding intervals to the rational expression, we obtain,

$x < 0$	$0 < x < \frac{3}{5}$	$x > \frac{3}{5}$
t.p. $x = -3$	t.p. $x = \frac{1}{5}$	t.p. $x = 3$
$\frac{3-5x}{x}$	$\frac{3-5x}{x}$	$\frac{3-5x}{x}$
$\frac{+}{-} = -$	$\frac{+}{+} = +$	$\frac{-}{+} = -$
< 0	> 0	< 0

(dividing points: 0, $\frac{3}{5}$)

i.e.

$$\frac{3-5x}{x} > 0 \Longleftrightarrow x \in \left(0, \frac{3}{5}\right)$$

(and

$$\frac{3-5x}{x} \leq 0 \Longleftrightarrow x \in (-\infty, 0) \cup [\frac{3}{5}, \infty)\ )$$

Notice that we must always *exclude* the values of x at which the rational expression is undefined (i.e. the 'extra' dividing points) from the solution set. This is only proper since even though these values are used in determining the sign of the rational expression in the corresponding intervals (i.e. on either side of these values), the expression is not defined at these specific values.

Example 3.8.9
Solve the inequality

$$-\frac{3}{x+1} < \frac{2}{x-4}$$

Solution
STEP 1: Make a zero on the right-hand side (avoid the temptation to cross-multiply)

$$-\frac{3}{x+1} - \frac{2}{x-4} < 0$$

STEP 2: Combine terms

$$\begin{aligned} \frac{-3(x-4)-2(x+1)}{(x+1)(x-4)} &< 0 \\ \iff \frac{10-5x}{(x+1)(x-4)} &< 0 \\ \iff \frac{5(2-x)}{(x+1)(x-4)} &< 0 \end{aligned}$$

STEP 3: Identify dividing points: $x = -1, 2, 4$.
Applying the test-point method we obtain:

$x < -1$	$-1 < x < 2$	$2 < x < 4$	$x > 4$
t.p. $x = -3$	t.p. $x = 0$	t.p. $x = 3$	t.p.$x = 5$
$\frac{5(2-x)}{(x+1)(x-4)}$	$\frac{5(2-x)}{(x+1)(x-4)}$	$\frac{5(2-x)}{(x+1)(x-4)}$	$\frac{5(2-x)}{(x+1)(x-4)}$
$\frac{+}{-\cdot-} = +$	$\frac{+}{+\cdot-} = -$	$\frac{-}{+\cdot-} = +$	$\frac{-}{+\cdot+} = -$
> 0	< 0	> 0	< 0

(Dividing points on the number line: -1, 2, 4.)

Hence,

$$-\frac{3}{x+1} < \frac{2}{x-4} \iff x \in (-1, 2) \cup (4, \infty) \tag{3.8.6}$$

(and

$$-\frac{3}{x+1} > \frac{2}{x-4} \iff x \in (-\infty, -1) \cup (2, 4)\ .)$$

Note that the values $x = -1, 4$ are once again excluded from the solution set - even if we were asked for

$$-\frac{3}{x+1} \leq \frac{2}{x-4}$$

we would include the point $x = 2$ in the solution set (3.8.6) but not $x = -1, 4$ where the original expression is undefined.

Exercises 3.8

1. Solve

   (i) $x - 2 < 5$ **(ii)** $-x + 2 \leq 3$ **(iii)** $2 \geq \dfrac{4 - 2x}{5} > -4$

2. Solve

   (i) $(x - 2)(x + 5) \leq 0$ **(ii)** $x^2 < 4x - 9$ **(iii)** $x^2 + 2x - 7 \leq 0$
   (iv) $(x + 3)x(x - 3) \geq 0$ **(v)** $(x + 1)^2(x - 1) > 0$

3. Solve

   (i) $\dfrac{x - 5}{x + 2} \leq 0$ **(ii)** $\dfrac{x - 1}{x^2 - 9} \geq 0$ **(iii)** $\dfrac{-3}{x + 1} < \dfrac{2}{x - 4}$
   (iv) $\dfrac{x + 4}{x^2 - 8x + 15} \geq 0$

3.9. ABSOLUTE VALUE

In this section, we review the basic properties of the *absolute value* and its use in the solution of algebraic equations and inequalities.

$|a|$, the *absolute value* of a, is the 'positive part' of the number a or the **distance** from a to the origin on the real line. In fact,

$$|a| = \begin{cases} a, & \text{if } a \geq 0, \\ -a, & \text{if } a > 0, \end{cases}$$

so that $|a|$ is (as required) always positive.

Properties

$$\begin{aligned} |a| &\geq 0 \quad \forall x \in R \\ |a| &= 0 \Longleftrightarrow a = 0 \\ |-a| &= |a| \\ |ab| &= |a|\,|b| \\ |a + b| &\leq |a| + |b| \text{ (triangle inequality)} \\ |a|^2 &= a^2 \end{aligned}$$

There is an important connection between absolute values and square roots:

$$\sqrt{a^2} = |a|$$
$$\text{e.g. } \sqrt{6^2} = |6| = 6, \quad \sqrt{(-6)^2} = |-6| = 6$$

In other words, the symbol $\sqrt{\cdot}$ is always taken to mean the positive square root (see Note 3.3.4) i.e. in general, for $x \in R$,

$$\sqrt{x^2} \neq x \quad \text{but, in fact,} \quad \sqrt{x^2} = |x|$$

Inequalities and the Absolute Value

$$|x| \leq a \quad \text{means} \quad -a \leq x \leq a$$
$$|x| \geq a \quad \text{means} \quad x \leq -a \quad \text{or} \quad x \geq a.$$

Example 3.9.1

(i) $|3x - 2| < 4$

$$\iff -4 < 3x - 2 < 4$$
$$\iff -2 < 3x < 6$$
$$\iff -\frac{2}{3} < x < 2$$
$$\text{i.e.} \quad x \in (-\frac{2}{3}, 2)$$

(ii) $|2x + 1| \geq 5$

$$\iff 2x + 1 \leq -5 \quad \text{or} \quad 2x + 1 \geq 5$$
$$\iff x \leq -3 \quad \text{or} \quad x \geq 2$$
$$\text{i.e.} \quad x \in (-\infty, -3] \cup [2, \infty)$$

(iii) $(x + 1)^2 \leq 4$

$$\iff |x + 1| \leq 2$$

take positive square root of both sides

$$\iff -2 \leq x + 1 \leq 2$$
$$\iff -3 \leq x \leq 1$$
$$\text{i.e.} \quad x \in [-3, 1]$$

Exercises 3.9

1. Solve

   (i) $|x+3| \geq 4$

   (ii) $|-x+2| < 3$

   (iii) $|6-4x| \geq |x-2|$ (Hint: square both sides)

   (iv) $(x-2)^2 > 16$

4. FUNCTIONS AND GRAPHS

Equations and inequalities represent relationships between variables. They can be thought of as algebraic objects or, by means of a coordinate system, given some geometric meaning. The relationship between algebra and geometry is very important. For example, when we seek information about an algebraic problem, it may be easier to look at the geometric interpretation and vice-versa (we shall see some examples of this in §4.2). First we review two different types of coordinate systems used in beginning calculus: the Cartesian or rectangular coordinate system and the polar coordinate system.

4.1. COORDINATE SYSTEMS

A coordinate system represents a point in the plane by an ordered pair of numbers called *coordinates*. The Cartesian or rectangular coordinate system represents the positions of points using distances along two perpendicular axes (commonly referred to as the $x-$ and $y-axes$) i.e.

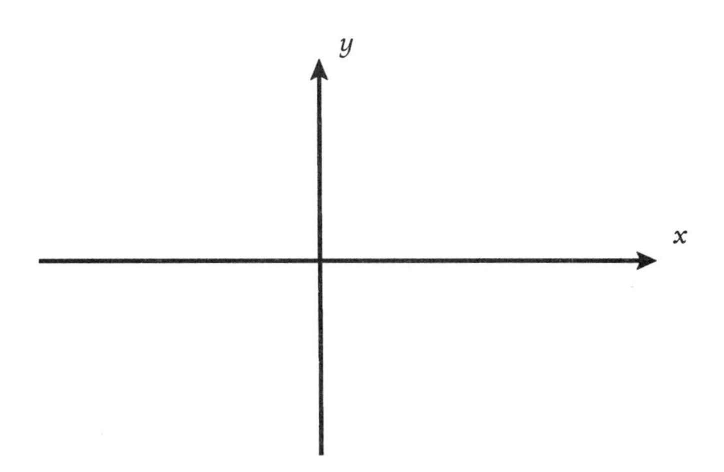

Recall that the set of real numbers (or points lying along the $x-axis$) is denoted by R. The generalization to the set of points (or pairs) (x, y) in the *Cartesian plane* is denoted by R^2 i.e.

$$R^2 = \{(x, y) : x, y \in R\}$$

The distance between any two points (x_1, y_1) and $(x_2, y_2)$is given by the *distance formula*,

$$d = \sqrt{(x_1 - x_2)^2 + (y_1 - y_2)^2} \tag{4.1.1}$$

For example, the distance between the points $(2,5)$ and $(-3,7)$ is (from (4.1.1)),

$$d = \sqrt{(2-(-3))^2 + (5-7)^2} = \sqrt{29}$$

The polar coordinate system introduced by Newton, is more convenient for many purposes. For example (see below), points in circular regions are easier to describe using polar rather than rectangular coordinates. In the polar coordinate system, the positions of points are described using an angle θ and the distance r along a radial line i.e.

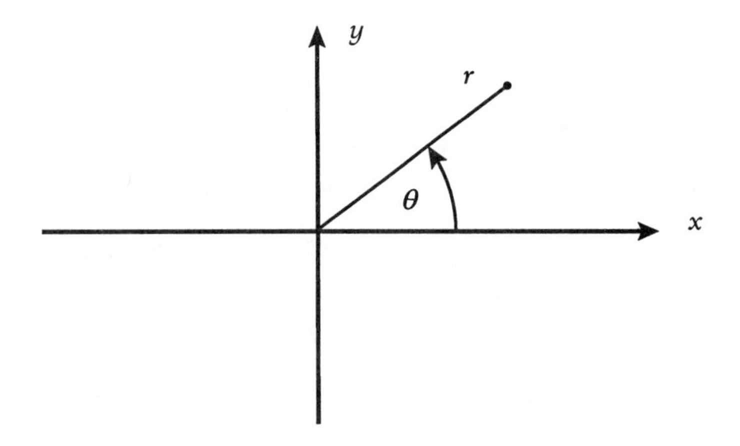

The connection between Cartesian and polar coordinates can be seen from the following diagram

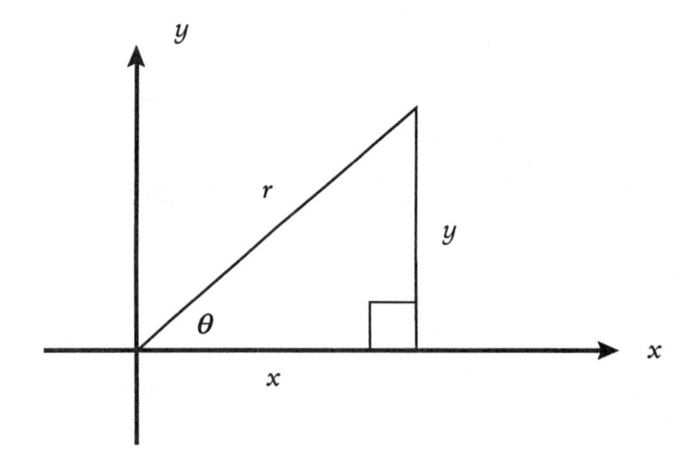

From the diagram, when r and θ are known we can obtain x and y using the transformation

$$x = r\cos\theta \qquad y = r\sin\theta \tag{4.1.2}$$

and when x and y are known we can obtain r and θ using the inverse transformation

$$r^2 = x^2 + y^2 \qquad \tan\theta = \frac{y}{x}$$

Example 4.1.1

Consider the following equations written in Cartesian form.

(i) $y = 3x + 2$ (ii) $x^2 + y^2 = 4$

The first represents the equation of a straight line while the second is the equation of a circle center $(0,0)$, radius 2 (see Note 4.1.2 below) - both in the Cartesian coordinate system. We can write these in polar form using (4.1.2) i.e.

$$\text{(i) } r\sin\theta = 3r\cos\theta + 2 \qquad \text{(ii) } r = 2$$

Clearly, in some cases, for example, (i), the representation becomes more complicated in polar form while in others, for example (ii), it simplifies. We can choose whichever coordinate system suits our purposes. The relationship between variables is not changed from one coordinate system to another - it is merely expressed in a different way e.g. $x^2 + y^2 = 4$ and $r = 2$ both represent the same geometric object - but with different algebraic representations!

Graphs

The geometric interpretation of an algebraic equation (in any coordinate system) is known as its *graph*. This is basically a plot of all *solution points* of the algebraic equation e.g. consider the equation

$$y = 3x + 2 \tag{4.1.3}$$

All solutions to this equation will take the form of pairs (x, y) or coordinates which satisfy the equation e.g. $(1, 5)$ is **one** (of course, there are many more!) solution of (4.1.3) since

$$5 = 3 \cdot 1 + 2$$

Collecting together all the solution points leads to the *solution set* of the equation and plotting this set in a coordinate system leads to the *graph* of the equation i.e.

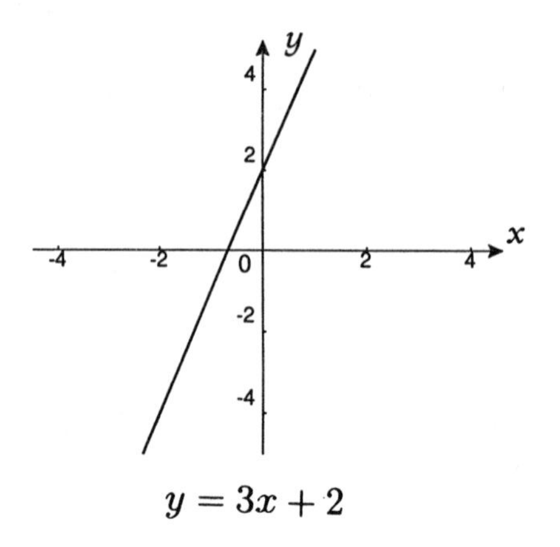

$y = 3x + 2$

Similarly, as noted above, the graph of the equation

$$x^2 + y^2 = 4 \text{ or } r = 2$$

takes the form of a circle, center $(0, 0)$, radius 2.

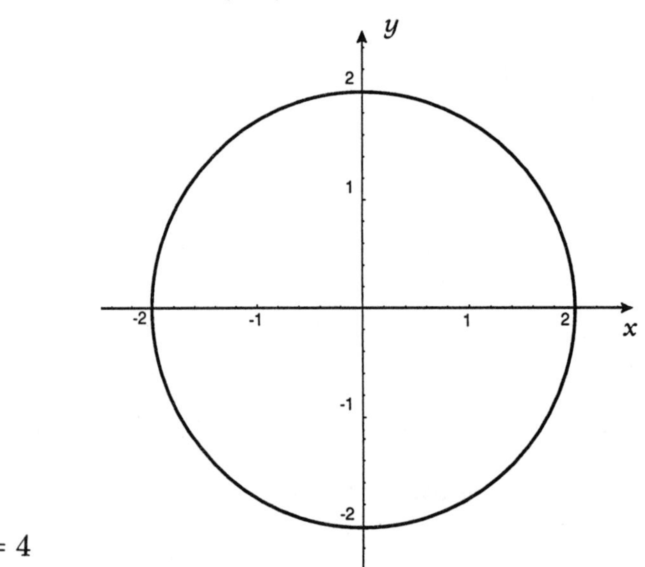

$r = 2$
or
$x^2 + y^2 = 4$

Note 4.1.2 - Conic Sections
The graphs of equations having the general form

$$ax^2 + bxy + cy^2 + dx + ey + f = 0 \qquad (a, b, c, d, e, f \text{ are constants}) \qquad (4.1.4)$$

are called *conic sections* since they all represent sections of a cone. Included in this set are straight lines, circles, ellipses, hyperbolas and parabolas. For example,

(i) If $b = 0$ and $a = c$, we have an equation of the form

$$(x-h)^2 + (y-k)^2 = r^2, \qquad h, k, r \text{ constants} \tag{4.1.5}$$

which is the general equation of a circle center (h, k) radius r.

(ii) If $a = b = c = 0$, we have the general equation of a straight line.

Often, we are given an equation in the form (4.1.4) and we are asked to identify its graph. In such cases, we use algebraic techniques from Section 3 to put the given equation in recognisable form:

Example 4.1.3
Consider the equation

$$x^2 + y^2 - 6y + 16x = 8$$

By completing the square in both x and y (see Note 3.7.5), we obtain

$$\begin{aligned} (x+8)^2 - 64 + (y-3)^2 - 9 &= 8 \\ \text{i.e. } (x+8)^2 + (y-3)^2 &= 81 \end{aligned}$$

which, when compared to (4.1.5), is easily seen to be the equation of a circle, center $(-8, 3)$, radius 9.

Exercises 4.1

1. Determine whether the three points $(-2, 8)$, $(1, 3)$ and $(2, 7)$ are the vertices of a right triangle.
2. Find an equation of the circle with center $(-1, 4)$ and radius 5.
3. Identify the conic section with equation

$$x^2 + 6x + y^2 + 4y + 9 = 0$$

4.2. EQUATION OF A LINE

In this section we collect together some useful properties of the graph and the equation of a straight line.

Slope - a measure of the steepness or inclination of the line i.e. if (x_1, y_1) and (x_2, y_2) are any two points on a straight line:

$$slope = m = \frac{y_2 - y_1}{x_2 - x_1} \quad , x_1 \neq x_2 \tag{4.2.1}$$

(i) If the line is vertical $(x_1 = x_2)$ the slope is *undefined* and the line has equation

$$x = a$$

where a is the intersection of the line with the $x - axis$.

(ii) If the line is horizontal $(y_1 = y_2)$ the slope is *zero* and the line has equation

$$y = a$$

where a is the intersection of the line with the $y - axis$.

(iii) If the slope is positive (negative), the graph of the line will rise (fall) as x increases (decreases).

(iv) Two lines are parallel if and only if their slopes are equal.

(v) Two lines are perpendicular if and only if their slopes are negative reciprocals of one another i.e.

$$m_1 = -\frac{1}{m_2}$$

Equation of a line
Let (x_1, y_1) be a point on a line with slope m and y-intercept c:

$$\begin{aligned} \text{Point-slope equation} &: \quad y - y_1 = m(x - x_1) \\ \text{Slope-intercept equation} &: \quad y = mx + c \end{aligned} \tag{4.2.2}$$

Example 4.2.1
Find the equation of the line passing through $(-1, -2)$ and $(2, 5)$.
Solution
To use either of the equations in (4.2.1), we must first find the slope m:

$$\text{From (4.2.1):} \quad m = \frac{5 - (-2)}{2 - (-1)} = \frac{7}{3}$$

(it doesn't matter which point you designate to be (x_1, y_1) when using (4.2.1) - the answer is always the same! Here we have used $(x_1, y_1) = (-1, -2)$ for convenience. Hence, from (4.2.2)

$$y + 2 = \frac{7}{3}(x + 1)$$

or

$$y = \frac{7}{3}x + \frac{1}{3}$$

In the next example we illustrate how the relationship between algebra and geometry allows us to deduce some very useful information regarding the (simultaneous) solution of algebraic equations representing straight lines (linear equations).

Example 4.2.2

Find any point of intersection of the lines

$$2x + 3y = 7 \text{ and } -x + y = 4 \qquad (4.2.3)$$

Solution

These are clearly equations of straight lines in the Cartesian plane. If we think about the various (geometric) possibilities, there are only three:

(i) the lines intersect once

or

(ii) the lines coincide

or

(iii) the lines never intersect i.e. they are parallel.

This gives us some very useful information regarding the solvability of the *simultaneous, linear* algebraic equations

$$\begin{aligned} ax + by &= c \\ dx + ey &= f, \qquad a, b, c, d, e, f \text{ constants} \end{aligned} \qquad (4.2.4)$$

The system (4.2.4) can have:

(i) one unique solution (x, y) - corresponding to **(i)** above.

or

(ii) an infinity of different solutions - corresponding to **(ii)** above (i.e. since the lines coincide there are an infinity of points common to both).

or

(iii) no solution - corresponding to **(iii)** above.

There are no other possibilities!

To find out which particular case applies to (4.2.3) we do some algebra. From the second of (4.2.3)

$$y = 4 + x$$

Substituting this into the first of (4.2.3) we obtain

$$2x + 3(4 + x) = 7$$

or

$$5x = -5$$

i.e.

$$x = -1 \text{ and, from either of (4.2.3), } y = 3$$

Hence, in this particular case, we have option **(i)** above i.e. a unique intersection point (solution). This is clear from the following diagram.

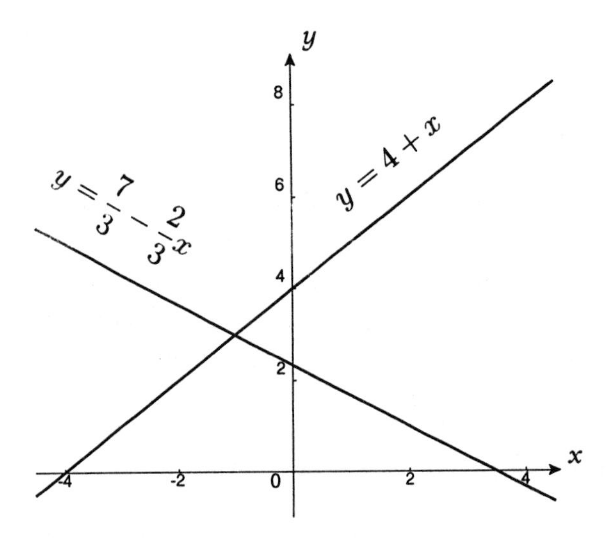

You may wish to verify that the following pairs are examples of options **(ii)** and **(iii)** above

$$\begin{aligned} x + y &= 3, \qquad 3x + 3y = 9 \qquad \text{(example of (ii) above)} \\ 2x - y &= 2, \qquad x - \frac{y}{2} = 2 \qquad \text{(example of (iii) above)} \end{aligned}$$

The study of equations of the form (4.2.4) constitutes a major part of the subject called *linear algebra*.

Exercises 4.2

1. Show that the slope of the line joining the point (x, x^2) and $((x+h), (x+h)^2)$ is $2x+h$ (note: this type of exercise is extremely common in the early stages of a beginning calculus course - when defining the slope of a curve *at a point* i.e. the *derivative of a function).*

2. Show that the slope of the line joining $(x, \frac{1}{x})$ to $(x+h, \frac{1}{x+h})$ is
$$-\frac{1}{x(x+h)}$$
($h \neq 0$ is constant in both cases)

3. Find the point of intersection (if any) of each of the following pairs of straight lines

   (a) $x - y = 7; \quad 2x + 3y = 1$

   (b) $4x - 6y = 7; \quad 6x - 9y = 12$

4.3. FUNCTIONS

Calculus is built almost entirely on the language of *functions.* For this reason, a good understanding of the relevant theory and applications is absolutely essential for success in calculus. Insufficient practice with the language of functions is arguably the single most common reason as to why students find difficulty in a beginning calculus course. For example, consider the following question (which always makes a good opener in a beginning calculus course).

$$\text{if } f(x) = 3 \text{ what is } f(x+h), \text{ where } h \text{ is constant ?} \tag{4.3.1}$$

As simple as this question may seem, there is always a significant number of students which returns the answer $3 + h$ (the correct answer is of course, 3) clearly demonstrating a lack of practice with the relevant concepts. Noting that the basic results from calculus are built almost entirely on a solid understanding of the theory of functions, you can begin to appreciate why students have so much difficulty in beginning calculus; it's like trying to build a house with little or no foundation - sooner or later the whole thing collapses!

With this in mind, we review the basics from the theory and applications of functions emphasising *practice* to achieve *fluency.*

Definition 4.3.1

A function is a *rule* which assigns to each element in one set (*domain*) exactly one value to another set (*range*).

Notation

A single letter like f (or g or h) is used to name the function. Then $f(x)$, read 'f of x' denotes the value that f assigns to x. e.g. if f denotes the squaring function,

$$\underbrace{f(x) = x^2}_{\text{formula for the function } f}$$

The formula for f tells us in a concise algebraic way what f does to any number.

IMPORTANT POINT**

The letter used for the domain variable (i.e. x above) is a matter of *no significance.* It is merely a *placeholder* (see §3.5) and as such, can be replaced with absolutely anything. It is there merely to make it easier for us to identify the action of the rule described by the function f. e.g.

$$\begin{aligned} f(x) &= x^2 \\ f(p) &= p^2 \\ f(h+1) &= (h+1)^2 \\ f(truck) &= (truck)^2 \\ f(\cdot) &= (\cdot)^2 \end{aligned}$$

all represent the same squaring function. The important point to remember is that once you have identified the placeholder, whenever you replace it with something on the left-hand side of the formula for the function, you must replace it with exactly the same thing wherever it appears on the right-hand side of the formula for the function. For example,

$$\begin{aligned} \text{if } f(x) &= x+3 \text{ then} \\ f(x+1) &= (x+1)+3 \\ f(2x) &= (2x)+3 \\ f(x+h) &= (x+h)+3 \\ f(\sqrt{x}) &= (\sqrt{x})+3 \\ f(x^2) &= (x^2)+3 \\ f(\theta) &= (\theta)+3 \\ f(\cdot) &= (\cdot)+3 \end{aligned}$$

Once you have practiced with this concept, you will feel much more comfortable with functions in general. Consider again the question posed in (4.3.1). The function is defined by

$$f(x) = 3 \tag{4.3.2}$$

This seems different in that there are no x's on the right-hand side of (4.3.2) -in the context of what was said above, this means that the result of the function does not change with the input i.e. it is a *constant function.* Hence,

$$\begin{aligned} f(2) &= 3 \\ f(\frac{1}{2}) &= 3 \\ f(x^2) &= 3 \\ f(x+h) &= 3 \\ f(\cdot) &= 3 \end{aligned}$$

This becomes even clearer when we consider the graph of the function defined by (4.3.2) i.e.

$$y = f(x) = 3$$

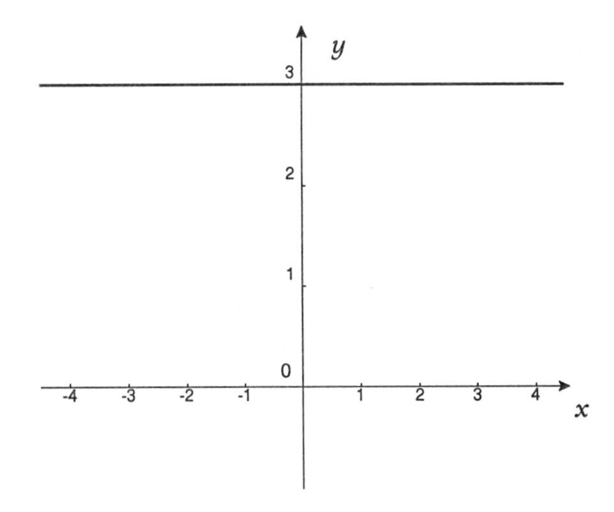

Clearly, every number in the domain of the function f is *mapped* to the single number 3 so that the range consists of only the number 3.

Example 4.3.2

(i) If $g(s) = s^3 - 1$, write an expression for

$$\frac{g(x+h) - g(x)}{h}, \quad h \neq 0$$

(ii) Let $h(\theta) = \sqrt{\theta + 1}$. What is $h(1)$, $h(\theta^2 + 1)$?

(iii) If $f(x) = x^2 + 2x + 1$, what is $f(2x + h)$, where h is constant ?

Solution

(i) $g(\cdot) = (\cdot)^3 - 1$. Hence,

$$g(x+h) = (x+h)^3 - 1$$

and

$$\frac{g(x+h)-g(x)}{h} = \frac{[(x+h)^3-1]-[x^3-1]}{h} = \frac{3x^2h+3xh^2+h^3}{h} = 3x^2+3xh+h^2$$

(ii) $h(\cdot) = \sqrt{(\cdot)+1}$. Hence,

$$h(1) = \sqrt{1+1} = \sqrt{2}$$
$$h(\theta^2+1) = \sqrt{(\theta^2+1)+1} = \sqrt{\theta^2+2}$$

(iii) $f(\cdot) = (\cdot)^2 + 2(\cdot) + 1$. Hence,

$$f(2x+h) = (2x+h)^2 + 2(2x+h) + \underset{\text{no } x \text{ dependency here!}}{1}$$
$$= 4x^2 + 4xh + h^2 + 4x + 2h + 1$$

Domain and Range

The *rule* of correspondence is the heart of a function, but a function is not completely defined until its *domain* is specified (the *range* then follows automatically). For example,

$$f(x) = x^2, \text{ domain } D = \{-2,-1,0,1,2,3\}$$

is essentially a different function than say,

$$f(x) = x^2, \text{ domain } D = R$$

If no domain is specified, it is understood to be the largest set of real numbers for which the rule for the function makes sense and gives real number values (c/f §3.3 - remember that in the real number system we are not allowed to divide by zero or take the even root of negative numbers). This is referred to as the *natural domain.* For example,

$$h(x) = \frac{1}{x-1} \text{ has natural domain } R \setminus \{1\}$$

(we exclude $x = 1$ from the domain since the result of the function does not make sense at $x = 1$ - just as in §3.3).

Example 4.3.3

Find natural domains for the functions defined by

(i) $f(x) = \dfrac{4x}{(x+2)(x-3)}$ **(ii)** $g(x) = \sqrt{x^2-4}$ **(iii)** $h(x) = \dfrac{1}{\sqrt{x^2-4}}$

Solution

(i) We have to 'remove' $x = -2, 3$ from the domain of f (since they lead to division by zero which is not allowed in the set of real numbers). Hence, the natural domain is

$$D = \{x \in R \setminus \{-2, 3\}\}$$

(ii) We cannot allow $x^2 - 4$ to become negative (else we end up taking the square root of a negative number which is not allowed in the set of real numbers). To exclude the 'offending' values of x, we first find where $x^2 - 4 \geq 0$ using the test-point method (§3.8).

$$x^2 - 4 = (x-2)(x+2) \geq 0$$

i.e.

$$x^2 - 4 \geq 0 \Longleftrightarrow x \in (-\infty, -2] \cup [2, \infty)$$

Hence, natural domain is

$$D = \{x \in (-\infty, -2] \cup [2, \infty)\} \tag{4.3.3}$$

(iii) Exactly as in **(ii)** above except that this time we also have to avoid division by zero (i.e. a combination of the problems faced in **(i)** and **(ii)** above). Hence, we exclude the values -2 and 2 from (4.3.3) and obtain the natural domain of h as

$$D = \{x \in (-\infty, -2) \cup (2, \infty)\}$$

As noted above, once the domain is understood and the rule is given, the range of the function is fixed - it is the set of 'mapped elements' e.g. in the above example,

(i) range is R since we can 'produce' any real number from this function.

(ii) range is R^+ i.e. non-negative real numbers (including zero) - we cannot 'produce' negative numbers from the positive square root function.

(iii) range is $R^+ \setminus \{0\}$ - same argument as in **(ii)** but this time we cannot produce the value zero from any element in the domain.

Sometimes we can use systematic methods to find the range e.g. consider the function defined by

$$f(x) = x^2 + 4x + 5$$

The (natural) domain is R (since there are no values of x which lead to division by zero or the even root of a negative number). If we write

$$x^2 + 4x + 5 = (x + 2)^2 + 1$$

it is clear that

$$f(x) \geq 1, \qquad \forall x \in R$$

Hence, the range is $[1, \infty)$.

Notice how the algebraic techniques from Section 3 are used over and over again. This again illustrates the cumulative nature of mathematics and the importance of a good *working* prerequisite.

Vertical lines test for functions

Not everything that looks like a function is actually a function - recall that there must be a unique value of $f(x)$ for each x in the domain of f if f is to define a function- a single domain element is not allowed to have more than one image e.g. if we propose the following rule for some $f(x)$,

$$y = f(x) \text{ where } y^2 = x \tag{4.3.4}$$

it is clear that for each x there are two y's or two image points. This is obvious from the graph of the equation

$$y^2 = x \tag{4.3.5}$$

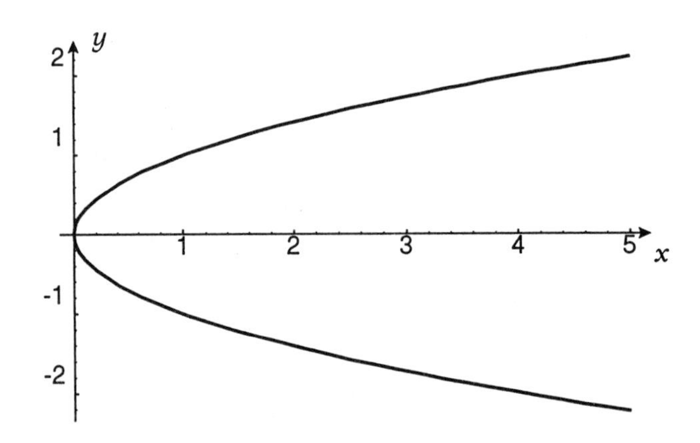

Hence, (4.3.4) does not define a function.

This procedure gives us a clue to devising a test to see if a given rule defines a function whenever we can sketch the graph of the equation defining the rule i.e.

> *a set of points in the Cartesian plane is the graph of a* ***function*** *if every vertical line in the plane intersects the graph no more than once.*

For example,

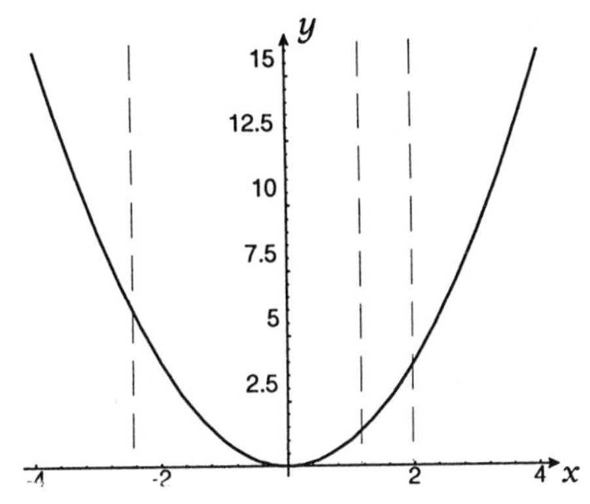

Vertical lines test on the graph of the function $y = f(x) = x^2$

Note 4.3.4

It is clear from the above graph of the equation (4.3.5) that we can 'make' a function from (4.3.5) by *restricting* the set of image points i.e. if we change the proposed rule to

$$y = f(x) \text{ where } y = \sqrt{x} \tag{4.3.6}$$

(remember that $\sqrt{\cdot}$ indicates the positive square root (see Note 3.3.4) so that (4.3.6) allows only positive image points) then this clearly defines a function with (natural) domain $D = R^+$ and range R^+ (including zero in both cases). This is clear from the graph of (4.3.6):

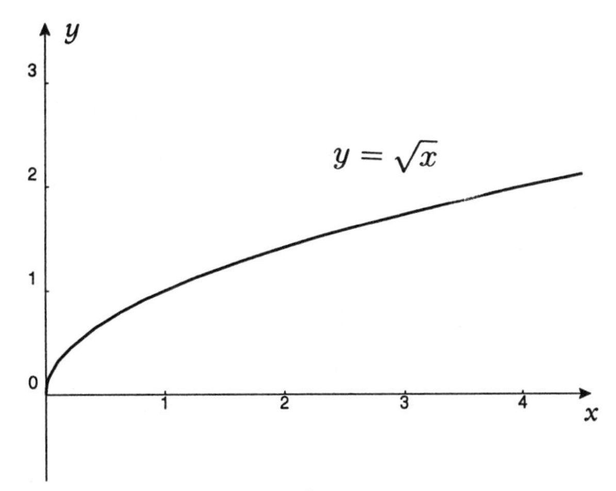

We end this section with a brief review of some relevant terminology.

Even or Odd Functions

The function $f(x)$ is said to be *even* if and only if

$$f(-x) = f(x)$$

and *odd* if and only if

$$f(-x) = -f(x)$$

For example,

$$\begin{aligned}
f(x) &= x^2 \text{ is even since } f(-x) = (-x)^2 = x^2 = f(x) \\
f(x) &= x^3 \text{ is odd since } f(-x) = (-x)^3 = -x^3 = -f(x) \\
f(x) &= x^2 + x^3 \text{ is neither even nor odd since } f(-x) = x^2 - x^3 \neq f(x) \text{ or } -f(x)
\end{aligned}$$

Note 4.3.5

Above, we used the concept of the *graph* of a function by simply assigning the $y-$*coordinate* to the value of the function f at the $x-$*coordinate*. Formally,

the graph of a function f is the set of points $\{(x, f(x)) : x \in \text{domain of } f\}$

e.g.

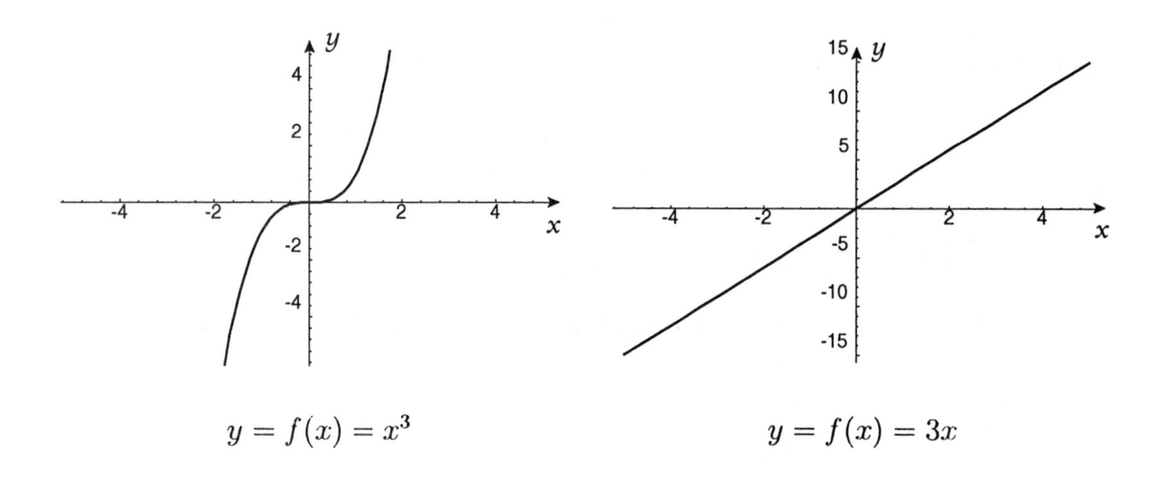

$y = f(x) = x^3$ $\qquad$ $y = f(x) = 3x$

Clearly,

(i) If $f(x)$ is even, its graph is symmetric about the $y-axis$.

(ii) If $f(x)$ is odd, its graph is symmetric about the origin i.e. given the graph for $x \geq 0$, the remainder is obtained by rotating this portion through $180°$ see the graph of $f(x) = x^3$ in Note 4.3.5.

Exercises 4.3

1. For the given function, find the given values.

 (a) $f(x) = \dfrac{1}{1+x}$; $\quad f(0), \quad f(x^2), \quad f(-2), \quad f(\sqrt{x})$.

 (b) $h(z) = 1 + z + z^2$; $\quad h(0), \quad h(-4), \quad h(z^5), \quad h(1+u^2)$.

(c) $g(t) = \sqrt{t+1}$; $g(0)$, $g(n^3 - 1)$, $g(\frac{1}{t})$.

2. Find the natural domain.

   (a) $f(x) = x^2 - 4$ (b) $f(t) = \dfrac{1}{t^2 - 1}$ (c) $F(u) = \dfrac{1}{\sqrt{u+1} - 2}$

3. Let $f(x) = \sqrt{x}$, $x \geq 0$. Show that for $h \neq 0$ (h is constant)

$$\frac{f(x+h) - f(x)}{h} = \frac{1}{\sqrt{x+h} + \sqrt{x}}$$

   (hint - rationalize the numerator) - this is exactly the type of calculation you will be required to perform in the early stages of your calculus course.

4. Let $f(x) = \dfrac{|x|}{x}$. Show that

$$f(x) = \begin{cases} 1, & x > 0, \\ -1, & x < 0. \end{cases}$$

   Find the domain and range of f.

5. Determine whether the given function is even, odd or neither

   (a) $f(x) = x^2 + 5$
   (b) $g(x) = x^3 + x^2 + 2$
   (c) $h(x) = x^{\frac{3}{5}}$

4.4. OPERATIONS WITH FUNCTIONS

Recall that a function is basically a rule and a domain (set of 'sensible' input values). Whenever we perform algebraic operations on functions e.g. add, subtract or multiply two functions, we obtain a new rule *and* a new domain. The new rule is usually straightforward to obtain but the new domain requires a little more attention e.g. 'troublesome' values which were not present in the original functions may have to be excluded from the new domain.

Example 4.4.1

Let $f(x) = \sqrt{1+x}$ and $g(x) = 4 - x^2$.

(i) The rule for the function

$$h(x) = f(x) + g(x)$$

is described by

$$h(x) = \sqrt{1+x} + 4 - x^2 \tag{4.4.1}$$

To obtain the (natural) domain of h, we look for the largest set of 'sensible' values for which (4.4.1) is true i.e.

$$D = \{x \in R : x \geq -1\}$$

(note that we could have obtained this result by noting that the domain of f is

$$D_1 = \{x \in R : x \geq -1\}$$

and the domain of g is $D_2 = R$ and the domain of h is the set

$$\{x : x \in D_1 and\ D_2\}$$

but it is often easier just to use (4.4.1) directly).

(ii) The rule for the function

$$h(x) = \frac{f(x)}{g(x)}$$

is described by

$$h(x) = \frac{\sqrt{1+x}}{4 - x^2} \tag{4.4.2}$$

The (natural) domain is again given by the largest set of values for which (4.4.2) makes sense i.e. to account for the square root we require $x \geq -1$ i.e. D_1 above. To account for possible division by zero we note that

$$4 - x^2 = (2 - x)(2 + x)$$

so that we must exclude $x = \pm 2$ from the domain of h. Hence, the domain of h is

$$D = \{x \in R : x \geq -1, x \neq 2\}$$

(the value $x = -2$ is already excluded by the requirement that $x \geq -1$). Note that again we could have proceeded formally and found D using

$$D = \{x : x \in D_1 and\ D_2 \text{ and } g(x) \neq 0\}$$

but using (4.4.2) directly is perhaps more logical relying less on formulae.

Similar arguments can be made to define the functions

$$h(x) = f(x) - g(x) \text{ and } h(x) = f(x) \cdot g(x)$$

Just remember that the new (natural) domain is again the largest set of values for which the new rule 'makes sense'.

Composite Functions

Consider the function

$$h(x) = \sqrt{x^2 + 1}$$

We can also think of this function as a composition of two other functions i.e.

$$h(x) = \sqrt{g(x)}$$

where

$$g(x) = x^2 + 1$$

or

$$h(x) = f(g(x)) = (f \circ g)(x) \tag{4.4.3}$$

where

$$f(s) = \sqrt{s} \quad (s \text{ is a 'dummy variable'})$$

(you can now see why it is necessary to be *fluent* in the language of functions to proceed to more advanced concepts!).

The function $f \circ g$ in (4.4.3) (read 'f on g') is called the *composition* of the functions f and g. Similarly, we can define

$$g \circ f = g(f(x)), \quad f \circ f = f(f(x)) \text{ and } g \circ g = g(g(x))$$

There are one or two important things to remember when working with compositions:

(i) The rule for e.g. $(f \circ g)(x) = f(g(x))$ is obtained in two stages:

Stage 1: g acts on an element x in its domain to produce $g(x)$.
Stage 2: f acts on the result from Stage 1 i.e. $g(x)$.

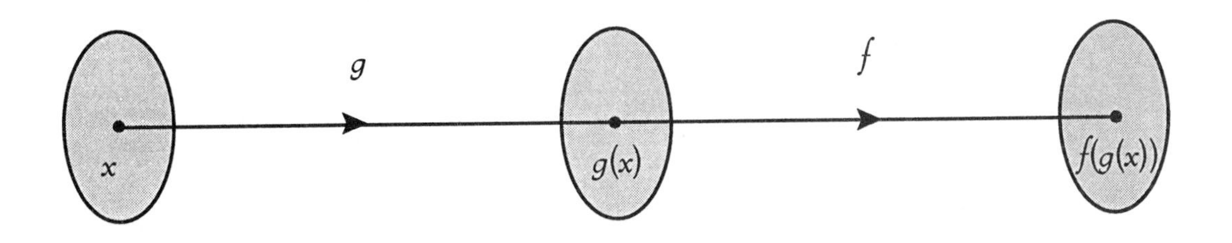

For example, if $f(s) = s^2$ and $g(x) = x + 1$, $\quad (f \circ g)(x) = f(x+1) = (x+1)^2$.

(ii) The domain for the composition is also in two stages e.g. for

$$(f \circ g)(x) = f(g(x))$$

Stage 1: x must be in the domain of g
Stage 2: $g(x)$ must be in the domain of f
i.e.
the domain of $f \circ g$ consists of those values of x such that both $g(x)$ and $(f \circ g)(x)$ are well-defined.

For example, if $f(\theta) = \theta^2 + 1$ and $g(s) = \sqrt{s}$,

$$(f \circ g)(s) = f(\sqrt{s}) = (\sqrt{s})^2 + 1 = s + 1 \qquad (4.4.4)$$

The domain of g is

$$D_1 = \{s \in R : s \geq 0\}$$

Next, from (4.4.4), $(f \circ g)(x)$ is well-defined for any $s \in R$. Hence, the domain of $f \circ g$ is

$$D = \{s \in R : s \geq 0\} \;= D_1 \cap R$$

i.e. just D_1.

Example 4.4.2
Let

$$f(x) = \frac{1}{\sqrt{x}}, \quad g(x) = x^2 - 4x$$

(i)

$$(f \circ g)(x) = f(g(x)) = f(x^2 - 4x) = \frac{1}{\sqrt{x^2 - 4x}} \qquad (4.4.5)$$

Now, the domain of g is R (since $g(x) = x^2 - 4x$ exists $\forall x \in R$). However, the expression for $(fog)(x)$ from (4.4.5) exists only for

$$x^2 - 4x > 0 \quad (\text{not } \geq 0 \text{ !!})$$
$$\text{i.e. } x(x-4) > 0$$

Using the test-point method (§3.8), we find that the expression for $(f \circ g)(x)$ in (4.4.5) exists only for

$$x \in (-\infty, 0) \cup (4, \infty)$$

Hence, the domain of $f \circ g$ is given by

$$D = \{ \underbrace{x \in R}_{x \text{ is in domain of } g} : \quad \underbrace{x \in (-\infty, 0) \cup (4, \infty)\}}_{(4.4.5) \text{ 'makes sense'}}$$

(ii)

$$(g \circ f)(x) = g(f(x)) = g(\frac{1}{\sqrt{x}}) = \frac{1}{x} - \frac{4}{\sqrt{x}} = \frac{\sqrt{x} - 4x}{x\sqrt{x}} \qquad (4.4.6)$$

Now, the domain of f is

$$D_1 = \{x \in R : x > 0\}.$$

The expression for $(gof)(x)$ from (4.4.6) exists only for

$$D_2 = \{x \in R : x > 0\}$$

Hence, the domain of $g \circ f$ is given by

$$D_1 \cap D_2 = D = \{\, x \in R : \underbrace{x > 0}_{x \in \text{domain of f and (4.4.6) 'makes sense'}} \}$$

(iii)

$$(f \circ f)(x) = f(f(x)) = f(\frac{1}{\sqrt{x}}) = \frac{1}{\sqrt{\frac{1}{\sqrt{x}}}}. = \frac{1}{\frac{1}{\sqrt{\sqrt{x}}}} = \sqrt{\sqrt{x}} = x^{\frac{1}{4}} \qquad (4.4.7)$$

Now, the domain of f is

$$D_1 = \{x \in R : x > 0\}$$

The expression for $(fof)(x)$ from (4.4.7) exists only for

$$D_2 = \{x \in R : x \geq 0\} \quad \text{(we can include 0 in this one)}$$

Hence, the domain of $f \circ f$ is given by

$$D_1 \cap D_2 = D = \{\, x \in R : \underbrace{x > 0}_{x \in \text{ domain of } f \text{ and (4.4.7) 'makes sense'}} \}$$

Note 4.4.3

The *decomposition* of functions is also extremely useful when, say, we wish to break down a complicated function (process) into several simpler ones e.g.

$$H(x) = \frac{1}{(x^2+3)^3} = (k \circ f \circ g \circ h)(x) = k(f(g(h(x))))$$

where

$$h(x) = x^2, \quad g(x) = x + 3, \quad f(x) = x^3 \text{ and } \; k(x) = \frac{1}{x}$$

The constituent parts are much simpler to analyse than the composition. Both composition and decomposition of functions will arise frequently in beginning calculus - particularly in the development of the rules for differentiation.

Exercises 4.4

1. Find $f+g$, $f\cdot g$, $\frac{f}{g}$, and their respective domains.

   (a) $f(x)=2x-5$, $g(x)=-4x$

   (b) $f(x)=\frac{x}{x+1}$, $g(x)=\frac{x-1}{x}$

2. Find $f\circ g$, $g\circ f$ and the domains of each.

   (a) $f(x)=x$, $g(x)=\frac{1}{2x}$

   (b) $f(x)=\sqrt{x+1}$, $g(x)=x^4$

   (c) $f(x)=\frac{x}{x+2}$, $g(x)=\frac{x-1}{x}$

3. Let

$$f(x)=2x+4,\quad g(x)=\frac{x}{2}-2.$$

   Show that

$$(f\circ g)(x)=(g\circ f)(x).$$

   (When this happens, we say that f and g are *inverse functions*).

5. POLYNOMIAL AND RATIONAL FUNCTIONS

Polynomial and rational functions are examples of *algebraic functions:* a function f is called algebraic if it can be constructed using algebraic operations (addition, subtraction, multiplication, division and taking roots) starting with polynomials. Here are two more examples:

$$f(x) = \sqrt{x+1}, \quad g(x) = \frac{x^3 - 16x}{x + \sqrt{x}} + (x - 23)\sqrt[4]{x^2 - 2}$$

Functions which are not algebraic are called *transcendental* (they *transcend* regular algebraic functions) e.g. trigonometric functions such as $\sin x$ and $\cos x$ (see §6) or logarithmic and exponential functions such as $\ln x$ or e^x (see §7) are transcendental functions. Despite their nature, transcendental functions are evaluated using polynomial representations (something you will discuss later in your calculus course). Polynomial functions are therefore extremely important as both the *building blocks* of all algebraic functions and as a basis for the numerical approximation of transcendental functions.

In this section we will survey some main results concerning polynomial functions and their graphs and then turn our attention to factoring higher order (greater than two) polynomials using long- or synthetic division and the theory of rational functions.

5.1. GRAPHS OF POLYNOMIAL FUNCTIONS

A polynomial function is a function which can be written as

$$p(x) = a_n x^n + a_{n-1} x^{n-1} + \cdots + a_2 x^2 + a_1 x + a_0 \tag{5.1.1}$$

where $a_n, a_{n-1}, \ldots, a_0$ are real constants and n is a positive integer e.g.

$$\begin{aligned} g(x) &= a_2 x^2 + a_1 x + a_0 \text{ is a quadratic function} \\ h(x) &= a_1 x + a_0 \text{ is a linear function} \end{aligned}$$

Basic Graphs

Consider the following graphs of the functions

$$f(x) = x, \quad g(x) = x^2 \text{ and } h(x) = x^3$$

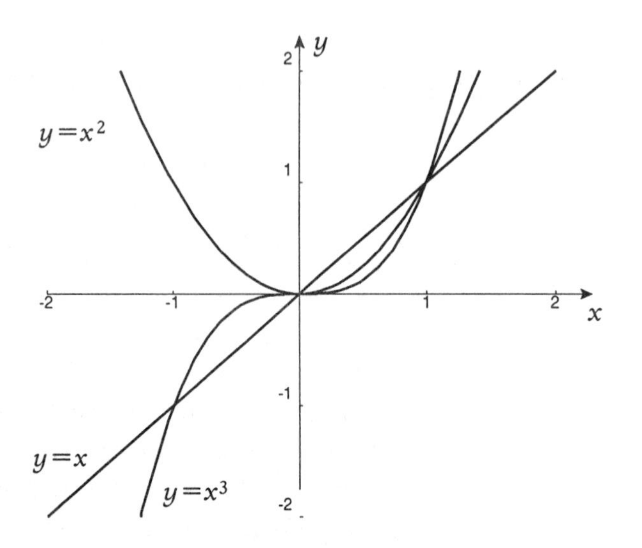

It is useful to note that the graphs of the functions

$$p(x) = x^{2n} \text{ and } q(x) = x^{2n+1}, \quad n \in N \tag{5.1.2}$$

closely resemble the graphs of

$$g(x) = x^2 \text{ and } h(x) = x^3$$

respectively. The main differences being

(i) the graphs of the functions (5.1.2) become flatter with increasing n for $x \in (-1, 1)$.

(ii) the graphs of the functions (5.1.2) become steeper with increasing n in the regions $x < -1$ and $x > 1$.

For example,

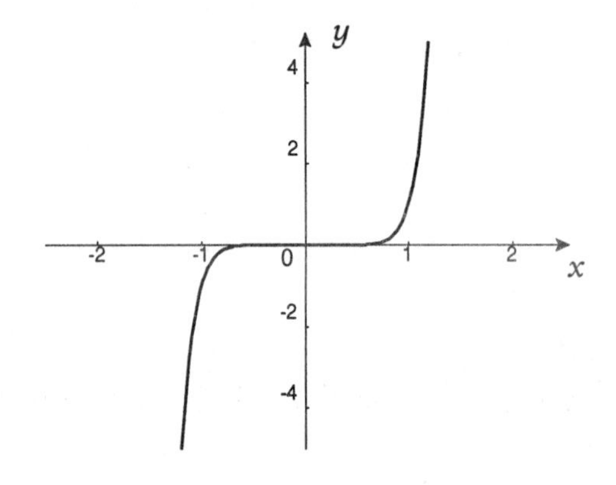

The graph of $y = x^9$

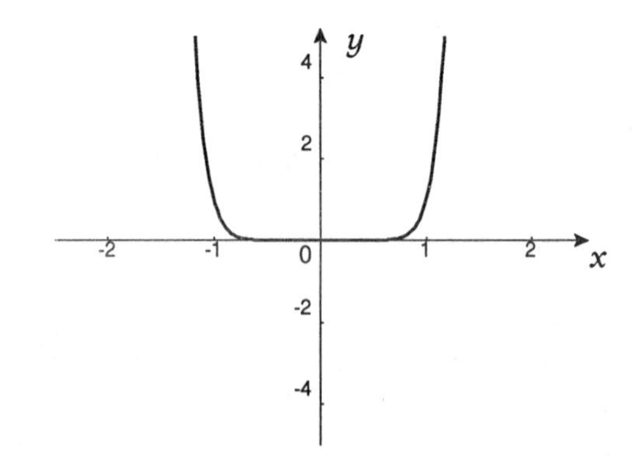

The graph of $y = x^{10}$

Quick Sketches

The following table shows how to obtain quick sketches of the graphs of functions obtained from a base function $y = f(x)$ through some combination of scaling and translation.

Table 5.1.1

Function	*Graph obtained from that of* $y = f(x)$ *by*
$y = -f(x)$	Reflection in the $x - axis$
$y = f(x - a)$	Shifting a units to the right
$y = f(x + a)$	Shifting a units to the left
$y - a = f(x)$	Shifting a units upward
$y + a = f(x)$	Shifting a units downward
$y = \|f(x)\|$	Reflection in the $x - axis$ of that part of the graph lying below the $x - axis$

Example 5.1.2

Consider the following graphs of the functions

$$y = f(x) = x - 2 \text{ and } y = h(x) = |x|$$

obtained using Table 5.1.1.

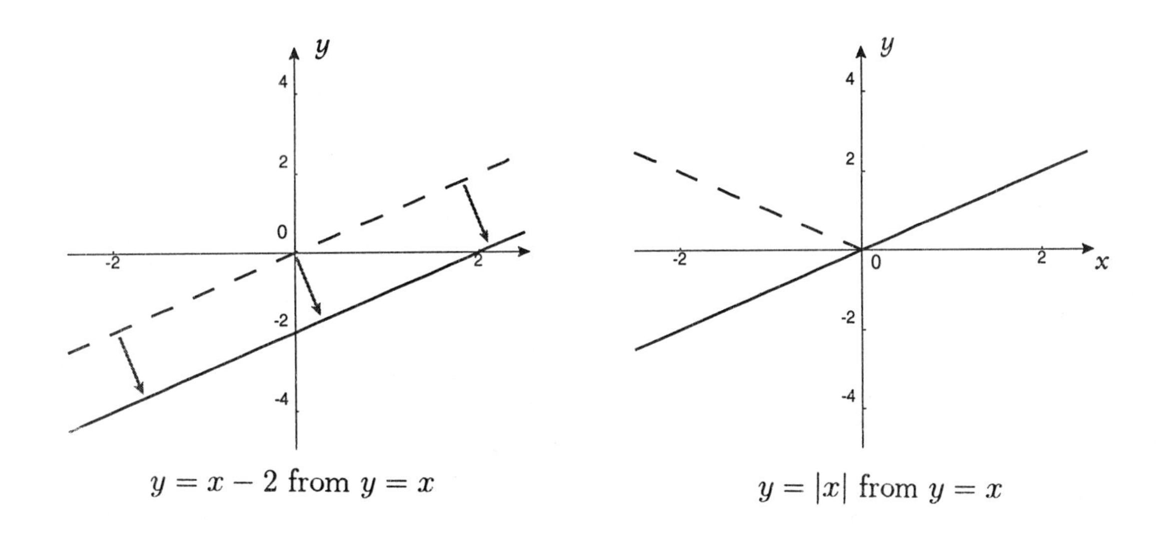

$y = x - 2$ from $y = x$

$y = |x|$ from $y = x$

Facts about real zeros of a polynomial

In §3.7 we introduced the term *zero* as being the solution of some quadratic equation. We also noted how the zeros led to factors of the quadratic and information on the corresponding graph. We are now in a position to extend this terminology to polynomial equations of any order.

For the polynomial function $p(x)$ given by (5.1.1) we define a *zero* or *root* as a number r such that

$$p(r) = 0$$

In addition, we have

Result 5.1.3
The following three statements are equivalent.

1. $p(r) = 0$ i.e. r is a zero of p.
2. $x - r$ is a factor of p.
3. $(r, 0)$ is an $x-$intercept of the graph of p.

For example,

$$\begin{aligned} \text{if } p(x) &= x^4 + 3x^3 - 8x^2 - 22x - 24 \\ p(3) &= 81 + 81 - 72 - 66 - 24 = 0 \\ p(-4) &= 256 - 192 - 128 + 88 - 24 = 0 \end{aligned}$$

Hence,

$$3, -4 \text{ are zeros of } p(x)$$

$$(x-3), \quad (x+4) \text{ are factors of } p(x)$$

and

$$x = 3, -4 \text{ are the } x - \text{ intercepts of the graph of } p(x)$$

In fact,

$$p(x) = (x-3)(x+4)(x^2+2x+2) \text{ (we will see how this is done in §5.3)}$$

and since $x^2 + 2x + 2$ has no real zeros (it's irreducible - §3.7), $x = 3, -4$ are the *only* two x-intercepts of the graph of $y = p(x)$.

Locating zeros of a polynomial function
The interplay between algebra and geometry is again useful in this time locating zeros (and hence factors - c/f Result 5.1.3) of a polynomial. Since zeros are $x-$intercepts of the corresponding graph, it follows that if the graph changes sign through some value of x then that value must be a zero. Formally,

Result 5.1.4
Let $p(x)$ be a polynomial function and suppose that $a < b$. If $p(a) < 0$ and $p(b) > 0$ (or vice-versa)

$$\exists \text{ at least one zero of } p(x) \text{ in the interval } (a, b)$$

This is clear from the following graphs.

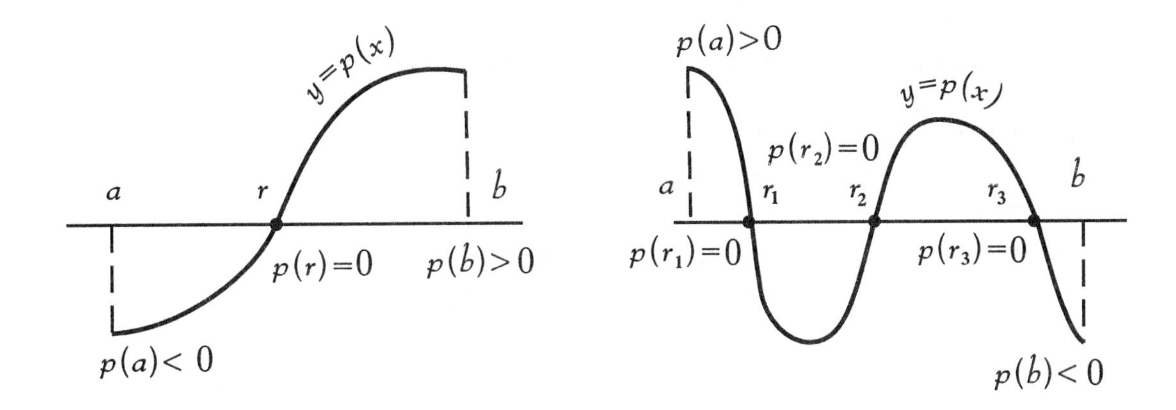

This gives us some idea of the location of the zero(s). Once we have this 'starting-point', we can then obtain a good approximation to the zero using some numerical method (e.g. Newton's Method - this will be discussed in your calculus course). An approximation to the exact value of the zero is sufficient for most applications. For example,

$$\begin{aligned} p(x) &= x^3 + x^2 + 7x - 3, \\ p(0) &= -3, \quad p(1) = 6 \end{aligned}$$

Hence, p has a zero between 0 and 1. In fact, using Newton's method, this zero has the approximate value 0.3971. There may be others but as we shall see later (§5.3), finding them for (i.e. factoring) polynomials of degree higher than 2 is not as straightforward as in the case of the quadratic.

Exercises 5.1

1. Find the zeros of $p(x)$.

   (a) $p(x) = (x-1)(x-2)(x-3)$

   (b) $p(x) = x(x-1)^2$

   (c) $p(x) = x^3 + 6x^2 + 8x$

   (d) $p(x) = x^3 - 9x^2$

2. Show that $p(x)$ has a zero in the given interval.

   (a) $p(x) = 4x^3 - 5x^2 + 4x - 7; \quad (1, 2)$

(b) $p(x) = 3x^5 + x^4 - 9x^2 + 3x - 4; \quad (1, 2)$

5.2. RATIONAL FUNCTIONS AND DIVISION OF POLYNOMIALS

Consider the cubic polynomial function given by

$$p(x) = x^3 - 6x^2 + 11x - 6$$

and suppose that we have noted that $p(1) = 0$. From the results established in §5.1 we immediately deduce that $(x - 1)$ is a factor of $p(x)$. How do we get the other factors? To answer this question, let us consider the real number analogy i.e.

$$6 = 2 \cdot \theta$$

To get the remaining factor θ, we simply divide 6 by the known factor 2 i.e.

$$\theta = \frac{6}{2} = 3$$

Returning now to the polynomial $p(x)$ above, it is clear that we can obtain the remaining factor(s) if we can perform the division

$$\frac{x^3 - 6x^2 + 11x - 6}{x - 1} \tag{5.2.1}$$

The process by which we arrive at a simplification of (5.2.1) is known as *long* or *synthetic* division.

Recall from §3.6 that the rational function

$$f(x) = \frac{p(x)}{q(x)}, \quad p, q, \text{ polynomials} \tag{5.2.2}$$

is called *proper* if degree $p <$ degree q and *improper* otherwise.

In this section we examine the process (long or synthetic division) by which we reduce the *improper* rational function (5.2.2) to the sum of a polynomial and a *proper* rational function i.e.

$$\underbrace{\frac{p(x)}{q(x)}}_{\text{improper}} = \underbrace{d(x)}_{\text{polynomial}} + \underbrace{\frac{r(x)}{q(x)}}_{\text{proper}} \tag{5.2.3}$$

Fortunately, long or synthetic division is a very systematic process.

Example 5.2.1
Consider the improper rational expression

$$\frac{4x^5 + x + 8}{3 - 2x + x^2}$$

Step 1: Arrange polynomials in order of descending powers of x -
(insert $0x^n$ as a placeholder if necessary)

$$\text{divisor} \longrightarrow x^2 - 2x + 3 \,\overline{\big)\, 4x^5 + 0x^4 + 0x^3 + 0x^2 + x + 8} \longleftarrow \text{ dividend}$$

Step 2: Divide the first term of the divisor into the first term of the dividend to get a quotient

$$\frac{4x^5}{x^2} = 4x^3 \longleftarrow \text{ quotient}$$

Step 3: Multiply the divisor by the quotient from Step 2

$$4x^3(x^2 - 2x + 3)$$

Step 4: Subtract the polynomial obtained in Step 3 from the dividend to obtain a remainder.

$$\begin{array}{rl} & 4x^3 \\ x^2 - 2x + 3 & \overline{\big)\, 4x^5 + 0x^4 + 0x^3 + 0x^2 + x + 8} \\ & \underline{4x^5 - 8x^4 + 12x^3} \\ & \quad 8x^4 - 12x^3 - 0x^2 \longleftarrow \text{ remainder} \end{array}$$

Step 5: Continue this process until the degree of the remainder is less than the degree of the divisor

$$\begin{array}{rl} & 4x^3 \\ \text{degree is } 2 \longrightarrow x^2 - 2x + 3 & \overline{\big)\, 4x^5 + 0x^4 + 0x^3 + 0x^2 + x + 8} \\ & \underline{4x^5 - 8x^4 + 12x^3} \\ & \quad 8x^4 - 12x^3 - 0x^2 \longleftarrow \text{ degree is } 4 \end{array}$$

Return to Step 2

The full calculation goes like this:

$$
\begin{array}{r|l}
 & 4x^3 + 8x^2 + 4x - 6 \\
\hline
x^2 - 2x + 3 & 4x^5 + 0x^4 + 0x^3 + 0x^2 + x + 8 \\
 & \underline{4x^5 - 8x^4 + 12x^3} \\
 & 8x^4 - 12x^3 \\
 & \underline{8x^4 - 16x^3 + 24x^2} \\
 & 4x^3 - 24x^2 + x \text{ (degree is 3 so continue)} \\
 & \underline{4x^3 - 8x^2 + 12x} \\
 & -16x^2 - 11x + 8 \text{ (degree is 2 so continue)} \\
 & \underline{-16x^2 + 32x - 48} \\
 & -43x + 56 \text{ (degree is 1 - STOP)}
\end{array}
$$

Hence,

$$\frac{4x^5 + x + 8}{x^2 - 2x + 3} = \underbrace{4x^3 + 8x^2 + 4x - 16}_{\text{polynomial}} + \underbrace{\frac{-43x + 56}{x^2 - 2x + 3}}_{\text{proper rational function}}$$

Example 5.2.2
Consider

$$\frac{p(x)}{q(x)} = \frac{3x^3 + x^2 - 15x - 5}{x + \frac{1}{3}}$$

Performing synthetic division:

$$
\begin{array}{r|l}
 & 3x^2 - 15 \\
\hline
x + \frac{1}{3} & 3x^3 + x^2 - 15x - 5 \\
 & \underline{3x^3 + x^2 + 0x} \\
 & -15x - 5 \\
 & \underline{-15x - 5} \\
 & 0
\end{array}
$$

Since the remainder is zero the division is exact and $(x + \frac{1}{3})$ is a factor of

$$p(x) = 3x^3 + x^2 - 15x - 5$$

That is,

$$\begin{aligned}
\frac{3x^3 + x^2 - 15x - 5}{(x + \frac{1}{3})} &= (3x^2 - 15) \\
\therefore\ 3x^3 + x^2 - 15x - 5 &= (x + \frac{1}{3})(3x^2 - 15) \\
&= (x + \frac{1}{3})3(x - \sqrt{5})(x + \sqrt{5})
\end{aligned}$$

Note 5.2.3
It is clear from (5.2.3) that if $q(x) = (x - c)$, we have

$$p(x) = (x - c)d(x) + r$$

where r is the *constant* remainder (since $\frac{r}{q}$ is proper and $q = x - c$ is of degree one so that r has to be constant by definition of a proper rational function). Hence,

$$p(c) = r \text{ is the remainder on dividing } p(x) \text{ by } (x - c)$$

That is, we can evaluate the remainder on dividing p by $(x - c)$ by simply finding $p(c)$ (this is known as the *Remainder Theorem*). For example, let

$$\frac{p(x)}{q(x)} = \frac{x^2 + 3x + 3}{x + 2} = x + 1 + \frac{1}{x + 2}$$

Then,

$$x^2 + 3x + 3 = (x + 2)(x + 1) + 1$$

With $c = -2$,

$$\text{remainder } = r = 1 = p(-2)$$

Further, in agreement with Result 5.1.3, if

$$p(c) = r = 0$$

we can write p as

$$p(x) = (x - c)d(x)$$

indicating that both $(x - c)$ and $d(x)$ are factors of p e.g. in Example 5.2.2 above,

$$p(c) = p(-\frac{1}{3}) = 0$$

so that we would have expected a zero remainder on division by $(x + \frac{1}{3})$.

Exercises 5.2

1. Find the quotient and the remainder when the first polynomial is divided by the second.

   (a) $x + 2$; $x + 1$

   (b) $4x^2 - 2x + 1$; $2x - 3$

   (c) $x^3 - x^2 + x + 3$; $x^2 - 2x + 3$

   (d) $x^3 - 1$; $x^2 + x + 1$

5.3. FACTORING HIGHER ORDER POLYNOMIALS

Suppose we wish to solve the polynomial equation

$$p(x) = 0 \tag{5.3.1}$$

where $p(x)$ is the nth degree polynomial (5.1.1) (by Result 5.1.3, this is equivalent to factoring the polynomial $p(x)$). We saw in §3.7 that if $p(x)$ is quadratic in x, there is a simple formula leading to the solutions of (5.3.1). For third-and fourth-degree polynomials $p(x)$, there are analogous formulae but they are extremely complicated, rendering them almost useless for practical purposes! If p is a polynomial of degree 5 or higher, it has been proved (by Niels Abel for the fifth degree polynomial in 1824 and later by Evariste Galois for polynomials of degree five and higher) that there is no such formula. Consequently, equations such as (5.3.1), where p is of degree 3 or higher, are usually investigated using numerical methods that give *approximations* to the solutions (and hence to the factors of p).

There are, however, some special cases where it is possible to factor $p(x)$ completely by first locating a single root (by e.g. trial and error - using Results 5.1.3 & 5.1.4) and then performing synthetic division. The most difficult part is locating the initial root! The process can be summarised as follows.

Step 1: Using Results 5.1.3 and 5.1.4 and trial and error, try to locate a single root of $p(x)$ i.e. try to find a value $x = r$ such that $p(r) = 0$.

Step 2: Once Step 1 has been completed, we have a factor, $(x - r)$of $p(x)$. Now divide $p(x)$ by $(x - r)$ to obtain the remaining factors.

Example 5.3.1
Consider the polynomial

$$p(x) = x^3 + 7x^2 + 17x + 14$$

Step 1: Try to find a factor with some 'intelligent' guesses:

$$\begin{aligned} p(0) &= 14 \text{ (no good!)}, \quad p(1) = 39 \text{ (no good!)} \\ p(-1) &= 3 \text{ (no good!)}, \quad p(-2) = 0 \text{ (OK!!!)} \end{aligned}$$

Hence, in agreement with Result 5.1.3 and Note 5.2.3, $(x+2)$ must be a factor of $p(x)$.

Step 2: Now divide $p(x)$ by $(x+2)$ to get a quadratic (the result must be a quadratic by Note 3.4.3 (i)) - remember to expect a zero remainder since $(x+2)$ is a factor:

$$\begin{array}{r|l} & x^2 + 5x + 7 \\ \hline x+2 & x^3 + 7x^2 + 17x + 14 \\ & \underline{x^3 + 2x^2} \\ & \quad 5x^2 + 17x \\ & \quad \underline{5x^2 + 10x} \\ & \qquad 7x + 14 \\ & \qquad \underline{7x + 14} \\ & \qquad\qquad 0 \text{ (as expected!)} \end{array}$$

Hence,

$$p(x) = x^3 + 7x^2 + 17x + 14 = (x+2)\underbrace{(x^2 + 5x + 7)}_{\text{IRREDUCIBLE}}$$

It is clear from the above example that we would save a lot of time if we could improve the process by which we 'guess' an initial root in Step 1. In fact, the following result helps us to 'pick our guesses' in a more organised way.

Result 5.3.2
Given a polynomial

$$p(x) = a_n x^n + a_{n-1} x^{n-1} + \cdots + a_1 x + a_0 \tag{5.3.2}$$

where $a_n, a_{n-1}, \ldots a_1, a_0$ are *integers*, $a_n \neq 0$,the *rational* roots are contained in the set of numbers formed by

$$\frac{\text{factors of } a_0}{\text{factors of } a_n}$$

Warning!

If none of these numbers is a root of $p(x)$ then $p(x)$ has no *rational* roots - but it may have *irrational* roots!!

Example 5.3.3

$$p(x) = 3x^3 - x^2 - 6x + 2$$

Possible *rational* roots:

$$x = \frac{m}{k} \qquad \text{where } m = \pm 1, \ \pm 2, \qquad k = \pm 1, \ \pm 3$$

In fact, $p(\frac{1}{3}) = 0$ (all the rest give $p(x) \neq 0$). Hence, $x = \frac{1}{3}$ is the only rational root and $(x - \frac{1}{3})$ is a factor. Synthetic division now gives

$$\begin{aligned} p(x) &= 3x^3 - x^2 - 6x + 2 = 3(x - \frac{1}{3}) \ \underbrace{(x^2 - 2)}_{(x-\sqrt{2})(x+\sqrt{2})} \\ &= (3x - 1)(x - \sqrt{2})(x + \sqrt{2}) \end{aligned}$$

Note that as well as the one rational root $x = \frac{1}{3}$, we have two irrational roots $x = \pm\sqrt{2}$ (which are not predicted by Result 5.3.2 but are nevertheless obtained through a process initiated by Result 5.3.2).

Example 5.3.4

(i)

$$p(x) = 2x^3 - \frac{11}{3}x^2 - x + \frac{2}{3}$$

To apply Result 5.3.1 we need *integer* coefficients! To this end, write

$$p(x) = \frac{1}{3}(6x^3 - 11x^2 - 3x + 2) = \frac{1}{3}p_1(x)$$

Possible rational roots of p_1:

$$x = \frac{m}{k} \qquad \text{where } m = \pm 1, \ \pm 2, \qquad k = \pm 1, \ \pm 2, \ \pm 3, \ \pm 6$$

Hence,

$$x = \pm 1, \ \pm 2, \ \pm\frac{1}{2}, \ \pm\frac{1}{3}, \ \pm\frac{2}{3}, \ \pm\frac{1}{6}$$

are possible rational roots. In fact,

$$p(2) = p(\frac{1}{3}) = p(-\frac{1}{2}) = 0$$

so that all the roots are rational! Hence, we have all (3) factors (although we could have obtained two from synthetic division once the first root had been identified),

$$(x-2),\ (x-\frac{1}{3}),\ (x+\frac{1}{2})$$

Thus,

$$p(x) = \frac{1}{3}(6x^3 - 11x^2 - 3x + 2) = \frac{1}{3}(x-2)(x-\frac{1}{3})(x+\frac{1}{2})$$

(ii)

$$p(x) = x^6 - 3x^2 - 2$$

Possible rational roots:

$$x = \pm 1,\ \pm 2$$

None of these work!!! i.e. none of the roots are rational! However,

$$\begin{aligned} p(x) &= x^6 - 3x^2 - 2 \\ &= (x^2)^3 - 3(x^2) - 2 \quad \text{(a cubic in } x^2\text{!)} \\ &= u^3 - 3u - 2 \end{aligned}$$

where

$$u = x^2$$

Now, examining the cubic

$$u^3 - 3u - 2 \tag{5.3.3}$$

we find that it has rational roots $u = -1, 2$ so that (5.3.3) has the factors

$$(u+1),\ (u-2)$$

Synthetic division gives us the final factor $(u+1)$. Hence, (5.3.3) has the roots $u = -1$ (twice) and $u = 2$. Thus,

$$u^3 - 3u - 2 = (u+1)^2(u-2)$$

and

$$\begin{aligned} p(x) = x^6 - 3x^2 - 2 &= (x^2+1)^2(x^2-2) \\ &= (x^2+1)^2(x-\sqrt{2})(x+\sqrt{2}) \end{aligned}$$

Hence, in fact, the real roots of $p(x)$ are irrational and there are complex roots $x = \pm i$ (twice).

Exercises 5.3

1. Show that $x - c$ is a factor of $p(x)$ and factor completely.

   (a) $p(x) = 2x^3 - 7x^2 + 9x - 4; \quad c = 1.$

   (b) $p(x) = x^3 - 7x^2 + 16x - 12; \quad c = 3.$

2. Factor the given polynomial into linear factors. You should be able to do it by inspection.

   (a) $x^2 - 5x + 6$

   (b) $2x^2 - 14x + 24$

   (c) $x^4 - 5x^2 + 4$ (hint: let $u = x^2$)

   (d) $x^4 - 13x^2 + 36$

3. Solve

   (a) $x^3 - 3x^2 - x + 3 = 0$

   (b) $2x^3 + 3x^2 - 4x + 1 = 0$

   (c) $\frac{1}{3}x^3 - \frac{1}{2}x^2 - \frac{1}{6}x + \frac{1}{6} = 0$ (hint: clear fractions)

6. TRIGONOMETRY

In calculus, it is important to develop a fluency in the manipulation of trigonometric identities. For example, the expression

$$\frac{1-\cos^2\theta}{\cos^2\theta\tan\theta}$$

is simply

$$\tan\theta$$

in disguise! Clearly, *recognizing* that expressions such as the one above can be simplified and then *performing* the simplification can save valuable time and effort.

In this section we review some of the basic theory from trigonometric functions and the use of trigonometric identities. With relevance to beginning calculus in mind, the emphasis is on the development of *fluency* in the manipulation of identities rather than in the fundamentals of trigonometry.

6.1. TRIGONOMETRIC FUNCTIONS

There are two ways to define trigonometric functions: using triangles or circles. The definition using triangles i.e.

$$\sin\theta = \frac{O}{H}, \quad \cos\theta = \frac{A}{H}, \quad \tan\theta = \frac{O}{A}$$

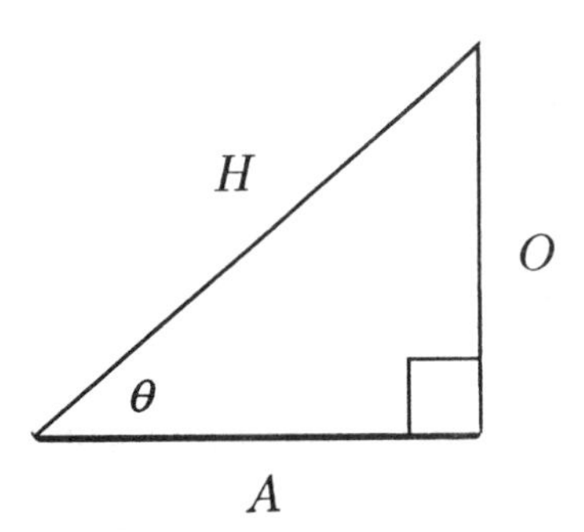

is too restrictive in that it treats only angles between 0°and 90°. The definition using circles on the other hand, leads to a much more general definition of the trigonometric functions since the angle can now take on any real value whatsoever i.e. going around the circle in one direction produces any positive value of the angle whilst changing direction gives us the corresponding negative values (see your calculus text for more information on the circle definition). This definition of the trigonometric functions is based on the concept of *radian measure*:

> The radian measure of an angle is equal to the numerical value of the length of the arc of the unit circle from the point (1,0) to the point (x, y)

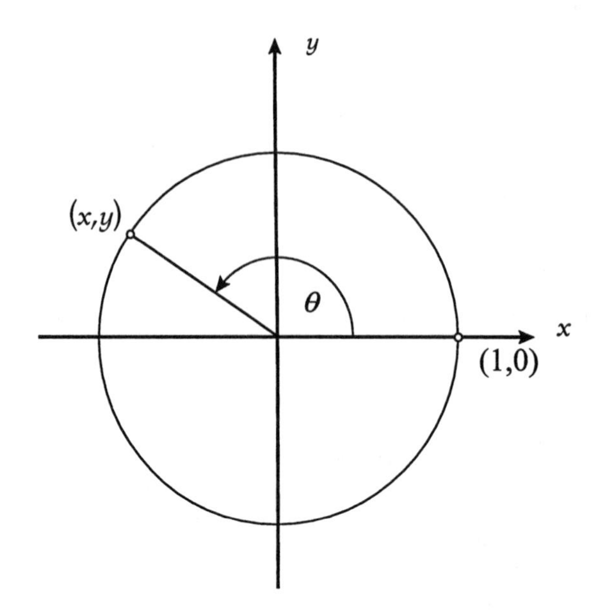

Hence,

$$360^\circ = 2\pi \text{ radians}$$

and to convert from degrees to radians (or radians to degrees) we multiply by $\frac{\pi}{180}$ (or $\frac{180}{\pi}$). In calculus we use radians to measure angles except where otherwise indicated. Hence, when we talk about the function f defined for all real numbers x by

$$f(x) = \sin x$$

it is understood that $\sin x$ means the sine of the angle whose *radian* measure is x. A similar convention holds for the other trigonometric functions $\cos, \tan, \csc, \sec$ and $\cot$.

Basic Properties of the Trigonometric Functions

We concentrate our attention on the functions sin, cos and tan (the others i.e. csc, sec and cot are defined in terms of sin, cos and tan so that their properties will follow).

Common Values:

x	0	$\frac{\pi}{6}$	$\frac{\pi}{4}$	$\frac{\pi}{3}$	$\frac{\pi}{2}$	π	$\frac{3\pi}{2}$	2π
$\cos x$	1	$\frac{\sqrt{3}}{2}$	$\frac{1}{\sqrt{2}}$	$\frac{1}{2}$	0	-1	0	1
$\sin x$	0	$\frac{1}{2}$	$\frac{1}{\sqrt{2}}$	$\frac{\sqrt{3}}{2}$	1	0	-1	0
$\tan x$	0	$\frac{1}{\sqrt{3}}$	1	$\sqrt{3}$	∞	0	∞	0

Furthermore,

$$\begin{aligned} -1 &\leq \cos x \leq 1 \\ -1 &\leq \sin x \leq 1 \\ \text{but note that } -\infty &< \tan x < \infty \qquad \forall x \in R \end{aligned}$$

The following diagram indicates which of $\sin x$, $\cos x$ and $\tan x$ is positive in each quadrant.

SINX	ALL
TANX	*COSX*

Periodicity

Both $\sin x$ and $\cos x$ are periodic with period 2π i.e. they repeat every 2π units. Hence,

$$\begin{aligned} \cos(x+2\pi) &= \cos x \\ \sin(x+2\pi) &= \sin x \end{aligned}$$

The function $\tan x$ on the other hand is periodic with period π. Hence,

$$\tan(x+\pi) = \tan x$$

Odd/Even

$$\begin{aligned} \cos(-x) &= \cos x : && \cos \text{ is an even function} \\ \sin(-x) &= -\sin x : && \sin \text{ is an odd function} \\ \tan(-x) &= -\tan x : && \tan \text{ is an odd function} \end{aligned}$$

Graphs

All of the above information (and more) is illustrated in the following graphs of the trigonometric functions.

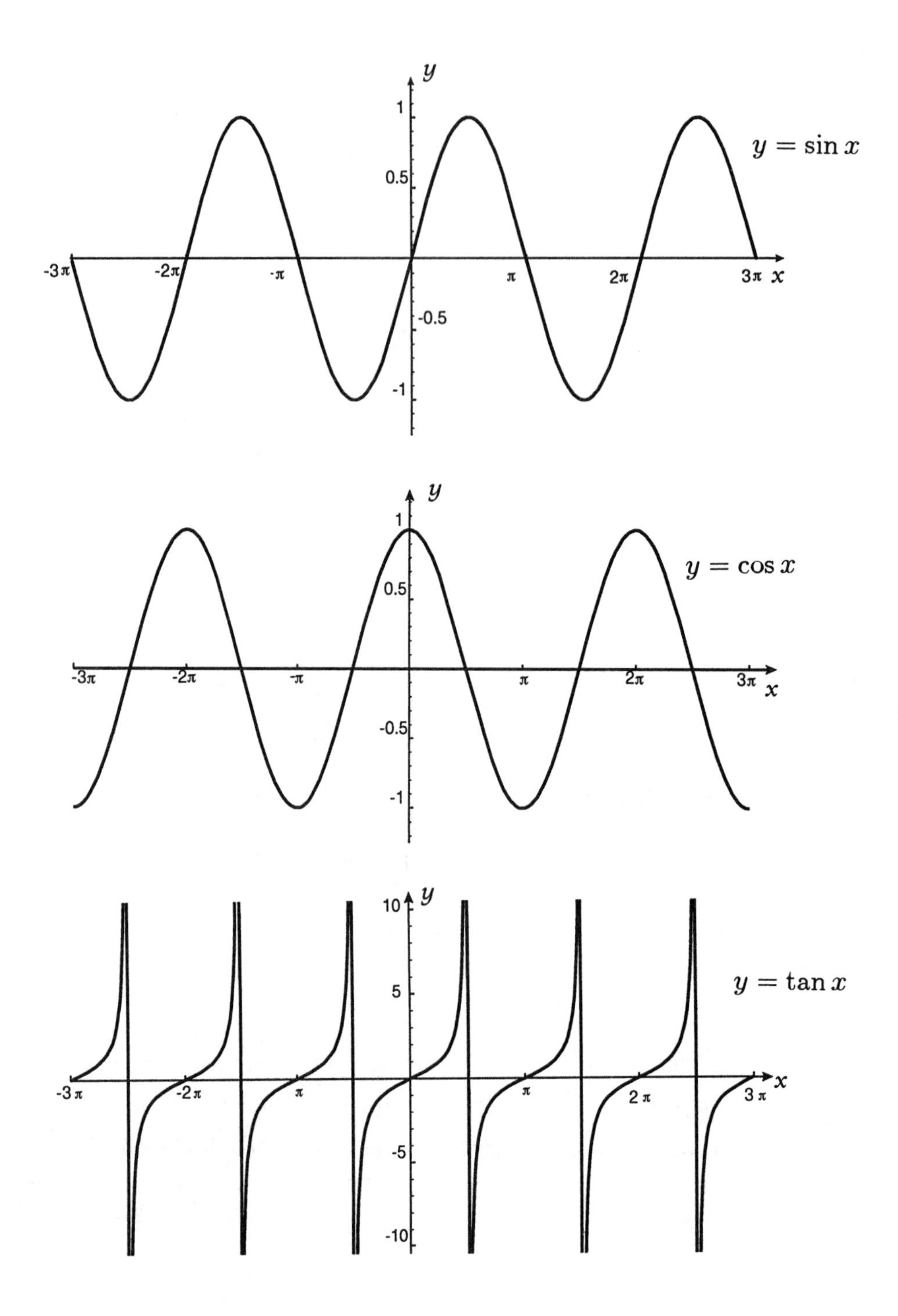

The lines $x = \ldots -\frac{3\pi}{2}, -\frac{\pi}{2}, \frac{\pi}{2}, \frac{3\pi}{2}, \ldots$ are called vertical asymptotes (the graph never touches them but bends towards them). They occur when $\tan x = \frac{\sin x}{\cos x}$ becomes undefined i.e. when $\cos x = 0$. You will encounter *asymptotes* frequently in your calculus course.

Other Trigonometric Functions

The tangent, cotangent, secant and cosecant of x are defined respectively as follows

$$\tan x = \frac{\sin x}{\cos x}, \qquad \cot x = \frac{\cos x}{\sin x}$$
$$\sec x = \frac{1}{\cos x}, \qquad \csc x = \frac{1}{\sin x}$$

Using these definitions and the above properties of $\sin x$, $\cos x$ and $\tan x$, it is not difficult to deduce the corresponding properties for $\cot x$, $\sec x$ and $\csc x$. Their graphs can be found in your calculus text.

Domains

$\sin\theta$, $\cos\theta$: domain is R

$\tan\theta$, $\sec\theta$: not defined when $\cos\theta = 0$

: $\therefore$ domain is $\{\theta \in R : \theta \neq \frac{\pi}{2} + n\pi,\ n = 0, \pm 1, \pm 2, \cdots\}$

$\cot\theta$, $\csc\theta$: not defined when $\sin\theta = 0$

: $\therefore$ domain is $\{\theta \in R : \theta \neq n\pi,\ n = 0, \pm 1, \pm 2, \cdots\}$

Clearly, there is much more that can be said about the trigonometric functions. However, as far as beginning calculus is concerned, the above represent the most important properties. In the next section, we use these properties to develop trigonometric identities for use in the simplification of expressions arising in calculus.

6.2. TRIGONOMETRIC IDENTITIES

The following is a comprehensive list of all the trigonometric formulae and identities you are ever likely to need in your calculus course. The formulae marked with an asterisk (*) should be committed to memory - they arise so frequently that you should really have them 'at your fingertips'. The rest can be referenced whenever necessary.

Table 6.2.1

1. *Definitions*

$$\tan x = \frac{\sin x}{\cos x}, \qquad \cot x = \frac{\cos x}{\sin x} = \frac{1}{\tan x}$$
$$\sec x = \frac{1}{\cos x}, \qquad \csc x = \frac{1}{\sin x}$$

2. *Pythagorean Identities*

$$\sin^2 x + \cos^2 x = 1, \quad 1 + \tan^2 x = \sec^2 x, \quad 1 + \cot^2 x = \csc^2 x$$

3. *Even-Odd Properties*

$$\begin{aligned} \cos(-x) &= \cos(x), \quad \sin(-x) = -\sin(x), \quad \tan(-x) = -\tan x, \\ \cot(-x) &= -\cot x, \quad \sec(-x) = \sec(x), \quad \csc(-x) = -\csc(x) \end{aligned}$$

4. *Reduction Identities*

$$\begin{aligned} \sin(x + 2\pi) &= \sin x, \quad \cos(x + 2\pi) = \cos x, \quad \tan(x + \pi) = \tan x \\ \sin(\frac{\pi}{2} - x) &= \cos x, \quad \cos(\frac{\pi}{2} - x) = \sin x, \\ \cos(x + \pi) &= -\cos x, \quad \sin(x + \pi) = -\sin x, \\ \cos(x - \pi) &= -\cos x, \quad \sin(\pi - x) = \sin x \end{aligned}$$

5. *Sum and Difference Identities*

$$\cos(x \pm y) = \cos x \cos y \mp \sin x \sin y$$
$$\sin(x \pm y) = \sin x \cos y \pm \cos x \sin y$$

$$\tan(x \pm y) = \frac{\tan x \pm \tan y}{1 \mp \tan x \tan y}$$

6. *Double-Angle Identities* *

$$\begin{aligned} \cos 2x &= \cos^2 x - \sin^2 x \\ &= 2\cos^2 x - 1 \\ &= 1 - 2\sin^2 x \end{aligned}$$

$$\sin 2x = 2 \sin x \cos x$$

7. *Product-to-Sum*

$$\begin{aligned} \sin x \sin y &= \frac{1}{2}[\cos(x - y) - \cos(x + y)] \\ \sin x \cos y &= \frac{1}{2}[\sin(x + y) + \sin(x - y)] \\ \cos x \cos y &= \frac{1}{2}[\cos(x + y) + \cos(x - y)] \end{aligned}$$

8. *Sum-to-Product*

$$\sin x + \sin y = 2\sin(\frac{x+y}{2})\cos(\frac{x-y}{2})$$
$$\sin x - \sin y = 2\cos(\frac{x+y}{2})\sin(\frac{x-y}{2})$$
$$\cos x + \cos y = 2\cos(\frac{x+y}{2})\cos(\frac{x-y}{2})$$
$$\cos x - \cos y = 2\sin(\frac{x+y}{2})\sin(\frac{y-x}{2})$$

Before we tackle some examples, we make two very important notes regarding the use of the above formulae.

(a) Learn to read the formulae properly i.e. using the notion of placeholder introduced in §3.4. Thus, the formula

$$\sin 2x = 2\sin x\cos x \tag{6.2.1}$$

is best read as

$$\sin 2(\cdot) = 2\sin(\cdot)\cos(\cdot) \tag{6.2.2}$$

so that (6.2.1) is not restricted to being a formula for $\sin 2x$ i.e. in the form (6.2.2) we can obtain the formula for the sine of twice any angle e.g.

$$\sin 4x = \sin 2(2x) = 2\sin 2x\cos 2x = 2[2\sin x\cos x]\cos 2x \text{ etc}$$

Similarly,

$$\sin 100x = \sin 2(50x) = 2\sin 50x\cos 50x \text{ etc}$$

(Note that the '2' on the RHS (right-hand side) of (6.2.2) does not change with the angle!)

(b) You must practice/'play' with the formula to achieve the desired level of *fluency* and familiarity (this also has the effect of exercising the 'mathematical part' of your brain!).

Example 6.2.2

(i) Simplify

$$y = (\tan x + \cot x)\sin x$$

Solution

Using Table 6.2.1,

$$\begin{aligned} y &= (\frac{\sin x}{\cos x} + \frac{\cos x}{\sin x}) \sin x \\ &= (\frac{\sin^2 x + \cos^2 x}{\sin x \cos x}) \sin x \\ &= \frac{1 \cdot \sin x}{\sin x \cos x} \\ &= \frac{1}{\cos x} \\ &= \sec x \end{aligned}$$

(ii) Show that

$$\frac{1 - \cos u}{\sin u} = \frac{\sin u}{1 + \cos u}$$

Solution

Avoid the temptation to cross-multiply! Instead, start with one side and show, using Table 6.2.1, that it reduces to the other side:

$$\begin{aligned} \text{Left-hand side} &= LHS = \frac{1 - \cos u}{\sin u} \\ &= \left(\frac{1 - \cos u}{\sin u}\right) \cdot \left(\frac{1 + \cos u}{1 + \cos u}\right) \quad \text{('rationalizing' the numerator)} \\ &= \frac{1 - \cos^2 u}{\sin u(1 + \cos u)} \\ &= \frac{\sin^2 u}{\sin u(1 + \cos u)} \\ &= \frac{\sin u}{1 + \cos u} \\ &= \text{RHS} \end{aligned}$$

as required.

Example 6.2.3

(i) Show that

$$\sin\theta = \frac{\tan\theta\cos\theta - \sin^2\theta}{1-\sin\theta}, \quad \theta \neq \frac{\pi}{2} + n\pi, \ n = 0, \pm 1, \ldots$$

Solution

There is not much that we can do with the left-hand side so we start with the right-hand side:

$$\begin{aligned} RHS &= \frac{\frac{\sin\theta}{\cos\theta}.\cos\theta - \sin^2\theta}{1-\sin\theta} \\ &= \frac{\sin\theta - \sin^2\theta}{1-\sin\theta} \\ &= \frac{\sin\theta(1-\sin\theta)}{1-\sin\theta} \\ &= \sin\theta \\ &= LHS \end{aligned}$$

as required.

(ii) Show that

$$\frac{1+\sin t}{1-\sin t} = (\sec t + \tan t)^2$$

Solution

$$\begin{aligned} \text{RHS} &= \left(\frac{1}{\cos t} + \frac{\sin t}{\cos t}\right)^2 \\ &= \left(\frac{1+\sin t}{\cos t}\right)^2 \\ &= \frac{(1+\sin t)^2}{\cos^2 t} \\ &= \frac{(1+\sin t)^2}{1-\sin^2 t} \end{aligned}$$

$$(\text{now note that } 1-\sin^2 t = (1-u^2) = (1-u)(1+u) \text{ where } u = \sin t\ )$$

$$
\begin{aligned}
&= \frac{(1+\sin t)^2}{(1-\sin t)(1+\sin t)} \\
&= \frac{1+\sin t}{1-\sin t} \\
&= \text{LHS}
\end{aligned}
$$

as required.

Note 6.2.4

As in the above example, when proving identities, it is always best to start with one side and reduce it to the other - rather than change the given identity.

Exercises 6.2

1. Prove the following identities

(a) $\dfrac{1-\cos^2 x}{\cos^2 x} = \tan^2 x$

(b) $(\cos x - \sin x)^2 = 1 - \sin 2x$

(c) $\dfrac{\cos^2\theta + 3\cos\theta + 2}{\sin^2\theta} = \dfrac{2+\cos\theta}{1-\cos\theta}$

(d) $\sqrt{\dfrac{1-\sin x}{1+\sin x}} = \dfrac{1-\sin x}{|\cos x|}$

1. Provide a counter-example to show that the following equation is not true for all values of x.

$$\sin x = \sqrt{\sin^2 x}$$

For which values of x is the equation valid ?

7. EXPONENTIAL AND LOGARITHMIC FUNCTIONS

Often, a beginning calculus course will include the applications of transcendental functions such as the exponential and logarithmic functions. Such a course would commonly be referred to as 'Calculus with Early Transcendentals'. There are many calculus textbooks designed specifically for this presentation.

This section is intended only for students taking the 'early transcendentals option'. In preparation for the latter, we review the basics from the theory of exponential and logarithmic functions.

7.1. EXPONENTIAL FUNCTIONS

The function

$$f(x) = b^x, \quad b > 0, \quad x \in R \tag{7.1.1}$$

is called *an exponential function of base* b and with *exponent* x.

Let $b = 2$. We have no trouble interpreting (7.1.1) when

(a) x is an integer e.g.

$$2^3 = 8, \quad 2^{-2} = \frac{1}{4}$$

or when

(b) x is rational e.g.

$$2^{\frac{2}{3}} = \left(\sqrt[3]{2}\right)^2$$

But what about when x is irrational ? i.e.

How do we interpret 2^{π} or $2^{\sqrt{2}}$?

One way might be to think of, for example, 2^{π} as a *limit* of a sequence of rational powers of 2 i.e.

$$2^3, 2^{3.1}, 2^{3.14}, 2^{3.141}, 2^{3.1415}, 2^{3.14159}, \ldots \longrightarrow 2^{\pi}$$

In fact, this method leads to a good approximation for 2^{π} ! It can also be applied to $f(x)$ in (7.1.1) for any irrational exponent x. Consequently, (7.1.1) 'makes sense' for *any* $x \in R$ (note that $b = 1$ gives $f(x) = 1$).

Note 7.1.1

From the definition (7.1.1), the exponential $y = b^x$ is never zero and in fact, always *positive* for any base b and exponent $x \in R$ i.e. $y = b^x > 0$, $\forall x \in R$.

Exponential functions are used to describe situations where the growth or decay of a quantity depends on the amount present e.g. population growth:

$$y = 100000 \cdot 2^{\frac{t}{5}}$$

represents the population of a country presently (t = time in years = 0) at 100000 and projected to double every 5 years

t	y
0	100000
5	2(100000)
10	$2^2(100000)$

Graphs of the exponential functions
The basic shape of the graph does not change with b - only the steepness:

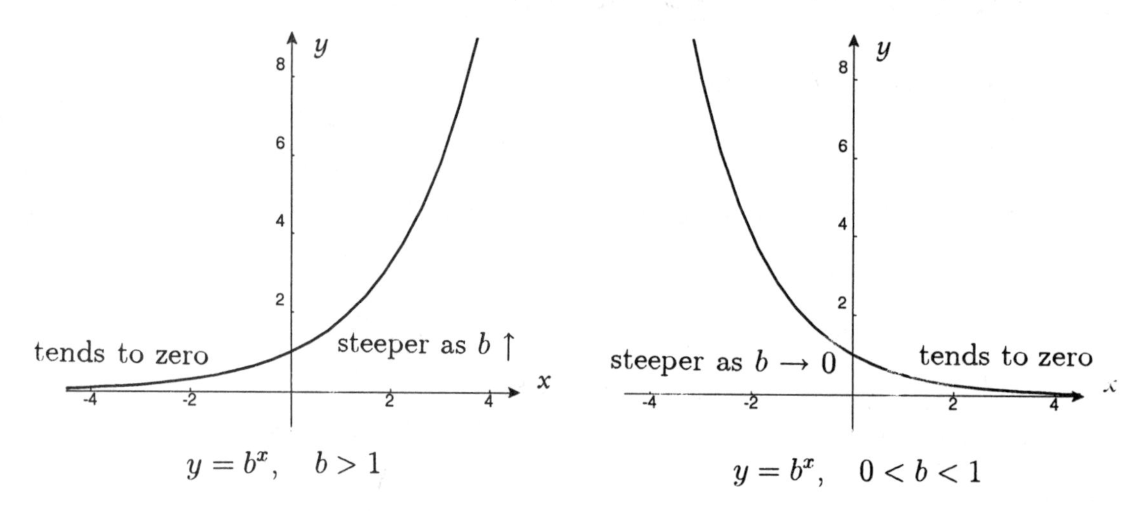

$y = b^x, \quad b > 1$ $\qquad\qquad$ $y = b^x, \quad 0 < b < 1$

Natural Exponential Function
The irrational number e (after Euler) defined as

$$e = \lim_{m \to \infty} \left(1 + \frac{1}{m}\right)^m = 2.7182818285\ldots$$

arises in a variety of applications (see your calculus text). When the base b in (7.1.1) takes on the value e, the function

$$y = f(x) = e^x, \quad x \in R.$$

is called the *Natural Exponential Function* (its name comes from the fact that it arises naturally in applications).

Properties of $f(x) = e^x$
Here we list the basic properties of the natural exponential function (but note that these properties hold for exponential functions (7.1.1) in general).

(i) $e^0 = 1$ **(ii)** $e^{(x+y)} = e^x \cdot e^y$ **(iii)** $e^{(x-y)} = \dfrac{e^x}{e^y}$

(iv) $\lim_{x \to \infty} e^x = \infty$ **(v)** $\lim_{x \to -\infty} e^x = 0$ **(vi)** $\lim_{x \to \infty} e^{-x} = 0$

(vii) $\lim_{x \to -\infty} e^{-x} = \infty$ **(viii)** $y = f(x) = e^x > 0 \quad \forall x \in R$

From Note 7.1.1 the natural exponential function

$$y = f(x) = e^x$$

has domain R but range R^+ i.e. $x \in R$ but $y \in (0, \infty)$.

The graphs of the functions

$$f(x) = e^x \text{ and } f(x) = e^{-x}$$

illustrate some of the above properties:

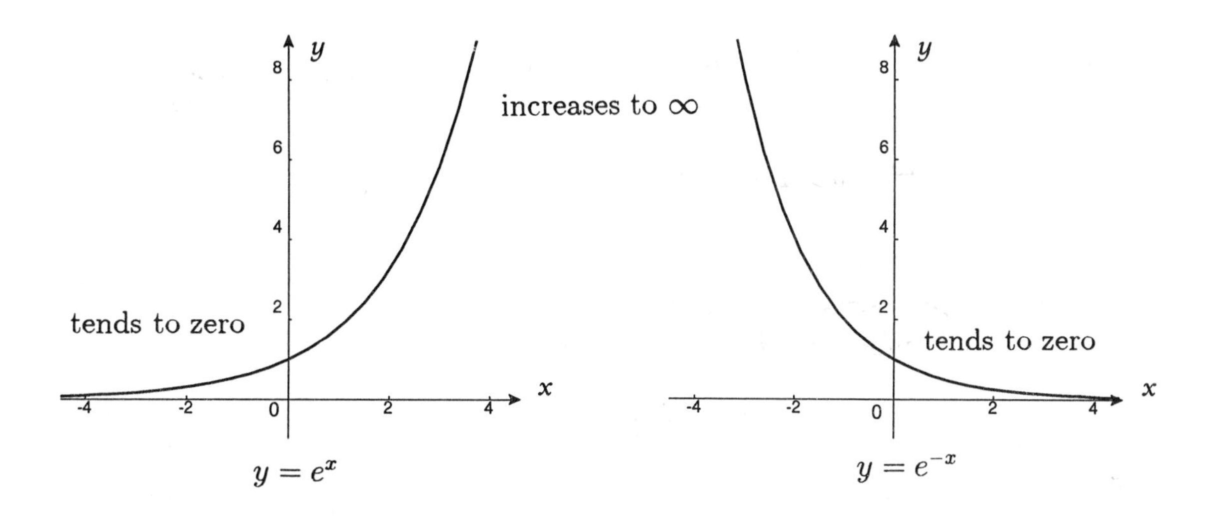

7.2. LOGARITHMIC FUNCTIONS

Suppose we know the exponential (base $b > 0$) of a certain number y. It is reasonable to ask if we can retrieve the exponent y from its exponential e.g.

$$\text{given } b^y, \quad \text{what is } y?$$

e.g. if $2^y = 16$, it is easy to see that $y = 4$. However, if $2^y = 17.215$, how then do we find y? To do this we introduce a new function called the *logarithm with base b*, denoted by $\log_b(\cdot)$, which acts as the *inverse* of the exponential function. It is defined as follows:

$$x = b^y \iff y = \log_b x \tag{7.2.1}$$

i.e. given the value $x = b^y$, the value of y is given by $y = \log_b x$.

Example 7.2.1

$$\left(\frac{1}{2}\right)^y = \frac{1}{8} \iff \log_{\frac{1}{2}}\left(\frac{1}{8}\right) = y$$

By inspection, it is clear that $y = 3$, hence,

$$\log_{\frac{1}{2}}\left(\frac{1}{8}\right) = 3$$

Note 7.2.2

From the definition (7.2.1) of the logarithm (see also Note 7.1.1) it is clear that for *any real value* y, x $(= b^y)$ has to be greater than zero at all times i.e.

$$\begin{aligned} \text{Domain of logarithmic function} &= \{x \in R : x > 0\} \\ \text{Range of logarithmic function} &= R \end{aligned}$$

Hence, we cannot take the logarithm of zero or any negative number although we can produce any number from a logarithm (this is 'the other way round' from the exponential function - which is what we would expect since the logarithm and the exponential are inverses of each other).

Graphs of the logarithmic functions

The following graphs of $y = f(x) = \log_b x$ are obtained by reflecting the graphs of the exponentials $y = b^x$ (see §7.1) in the line $y = x$ (i.e. switching the roles of x and y):

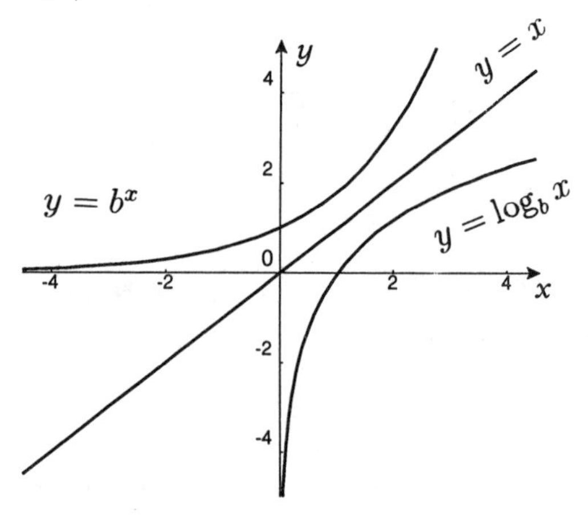

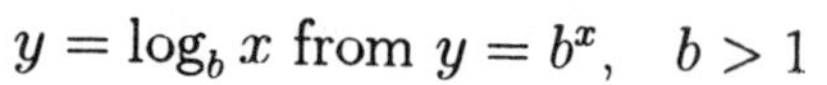

$y = \log_b x$ from $y = b^x$, $b > 1$

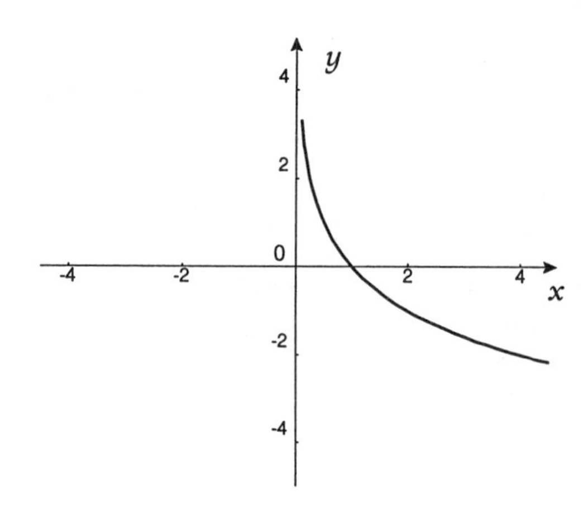

$y = \log_b x$, $0 < b < 1$

As with the graphs of the exponential functions with base $b > 0$, we can see that the basic shape of the graphs of the logarithmic functions do not change with a particular value of b in the given range - only the steepness.

Definition 7.2.3
Two of the most commonly used logarithmic functions are

(a) The *Common Logarithm:*

$$f(x) = \log x = \log_{10} x \quad \text{(i.e. base } b = 10)$$

and

(b) The *Natural Logarithm:*

$$f(x) = \ln x = \log_e x \quad \text{(i.e. base } b = e)$$

The following formulae is useful when changing the base of a logarithm from say b to a.

$$\log_a x = \frac{\log_b x}{\log_b a}$$

Properties of the natural logarithm

The following properties hold for logarithms of any base $b > 0$ but are listed specifically for the natural logarithm because of its importance:

(a) Domain $= \{x \in R : x > 0\}$, Range $= R$

i.e. $y = \ln x$ is defined *only for* $x > 0$

$$\begin{aligned} \ln 1 &= 0 \\ \ln e &= 1 \\ \ln(xy) &= \ln x + \ln y \\ \ln(\frac{x}{y}) &= \ln x - \ln y \\ \ln x^r &= r \ln x, \quad x, y > 0, \quad r \in R \end{aligned} \qquad (7.2.2)$$

(c)

$$\begin{aligned} \ln e^x &= x, \quad x \in R \\ e^{\ln x} &= x, \quad x > 0 \end{aligned}$$

(these two basically reflect the fact that the exponential and the logarithm are *inverses of* one another)

(d)

$$\lim_{x\to 0^+} \ln x = -\infty$$
$$\lim_{x\to\infty} \ln x = \infty$$

Some of these properties are reflected in the graph of the natural logarithm:

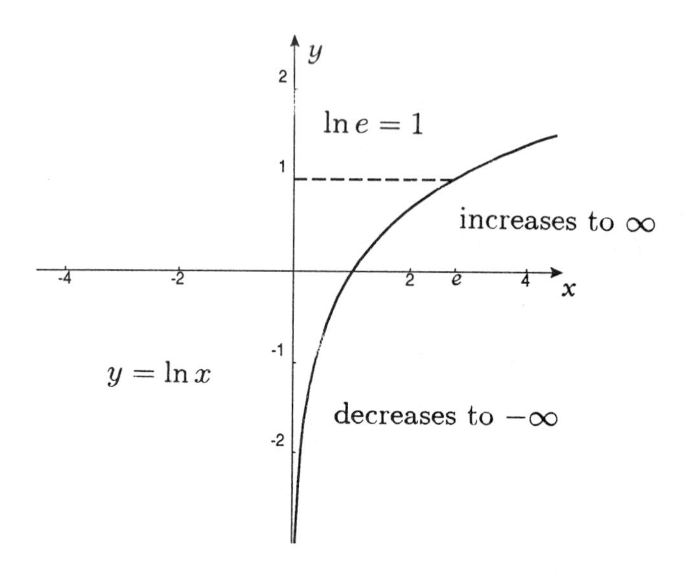

Example 7.2.4

(i) Simplify the expression

$$\ln\left[(x+y)^3(xy)^4\right]^{\frac{1}{2}}, \quad x, y > 0$$

Solution
From (7.2.2) we have

$$\begin{aligned}
\ln\left[(x+y)^3(xy)^4\right]^{\frac{1}{2}} &= \frac{1}{2}\ln\left[(x+y)^3(xy)^4\right] \\
&= \frac{1}{2}[\ln(x+y)^3 + \ln(xy)^4] \\
&= \frac{1}{2}[3\ln(x+y) + 4\ln(xy)] \\
&= \frac{1}{2}[3\ln(x+y) + 4\ln x + 4\ln y]
\end{aligned}$$

(ii) Simplify

$$2\ln(x+y) - \frac{1}{2}\ln(x-y), \quad x > y > 0$$

Solution
Using (7.2.2), we obtain

$$\begin{aligned} 2\ln(x+y) - \frac{1}{2}\ln(x-y) &= \ln(x+y)^2 - \ln(x \\ &= \ln\left[\frac{(x+y)^2}{\sqrt{x-y}}\right] \end{aligned}$$

Exercises 7.2

1. Simplify the following expressions.

   (a) $\ln(e^{\frac{1}{2}}e^{\frac{3}{2}})$

   (b) $e^{10\ln x}$, $x > 0$

   (c) $\ln\left(\frac{1}{e^{-x}}\right)$

   (d) $\ln(x^2 + 4x + 4)$

   (e) $4\ln\sqrt{x} + 3\ln(x^{\frac{1}{3}})$, $x > 0$

2. Solve for x

$$2^{x-2} = 3^x$$

3. Find the domain of the function

$$f(x) = \ln\frac{x}{4-x}$$

ANSWERS TO EXERCISES

3.3 Exponents and Roots

Q1. (a) $\dfrac{4a^4c^4}{b^5}$ (b) $\dfrac{4^4}{z^{28}}$ (c) $\dfrac{ab^2}{b^2+a}$ (d) $\dfrac{8y^{18}}{27x^9}$ (e) $\dfrac{6}{b^3d^3}$ (f) $\dfrac{xy}{y-x}$

Q2. (a) $x^{\frac{4}{5}}y^{\frac{6}{5}}$ (b) $3^{\frac{2}{3}}a^{\frac{2}{3}}b^{\frac{4}{3}}$ (c) $\left(x^{\frac{1}{2}}+y^{\frac{1}{2}}\right)^{\frac{1}{2}}$ (d) $xy^{-\frac{5}{4}}$ (e) x^4
(f) $\dfrac{b^{\frac{1}{2}}}{a}$

Q3. (a) $2(\sqrt{3}+\sqrt{2})$ (b) $\frac{1}{2}(\sqrt{5}-\sqrt{3})$ (c) $\dfrac{\sqrt{x}+\sqrt{a}}{x-a}$

3.4 Polynomials

Q1. (a) $12t^2-26t+10$ (b) y^2-2y+1 (c) $12x^3+17x^2+3x-2$
(d) $x^2+20x+100$ (e) x^2-64 (f) $4t^2-20t+25$ (g) $4x^8-25x^2$
(h) $t^6+2t^4+4t^3+t^2+4t+4$ (j) $-t^6+t^2+4t+4$ (k) $x^3+6x^2+12x+8$
(l) $27u^3+27u^2+9u+1$ (m) $u^3+12u^2v+48v^2u+64v^3$

Q2. (a) $\dfrac{x^2+2xy+y^2}{x^2y^2}$ (b) $x^2+y^2+z^2-2xy+2xz-2yz$
(c) $x^2-2+\dfrac{11}{x}-\dfrac{3}{x^2}+\dfrac{7}{x^3}+\dfrac{10}{x^4}$

Q3. -1

Q4. $\dfrac{5}{2}$

3.5 Factoring

Q1. (a) $x(x+5)$ (b) $y^2(y+4)$ (c) $8x^2y^3[1+2x^3y-3xy^3]$
(d) $(b-a)(a-2b)(5a+8b)$

Q2. (a) $(w-1)(w+1)$ (b) $(x-2y)(x+2y)$ (c) $(x^{50}-y^{25})(x^{50}+y^{25})$

Q3. (a) $(z-1)^2$ (b) $(3t+1)^2$ (c) $2(w+8)^2$

Q4. (a) $(t-2)(t-1)$ (b) $(z-8)(z+5)$ (c) $(xy-3z)^2$
(d) $(x-3y-z)(x-3y+z)$ (e) $(x+6)(x+7)$

(f) $3(x-2)^2$ (g) $(5x+2y)(7x+3y)$
(h) $(3x+y)(2x-5y)$ (j) $(x+2)(x^2-2x+4)$

Q5. (a) $(2x-3)(x^2-3x+39)$ (b) $(x+y+2)(x+y+1)$

Q6. (a) $5x^{-4}(2x^7+4x^5+3x^3-1)$
(b) $\frac{1}{2}(2x^2-x+1)^{-\frac{1}{2}}(x^3+1)^{-\frac{2}{3}}(8x^4-3x^3+2x^2+4x-1)$
(c)
$$\frac{-2(x^2+1)^2(4x^2-3x-2)}{(1-2x)^3}$$

3.6 Rational Expressions

Q1. (a) $\frac{1}{x-6}$, $x \neq \pm 6$ (b) $\frac{y}{5}$, $y \neq -1$ (c) $\frac{(x+2)^2}{(x-2)}$, $x \neq \pm 2$
(d) $\frac{2(4x-3)}{(x+2)(x-2)}$, $x \neq \pm 2$

Q2. (a) $\frac{x^2+10x-3}{(x-2)^2(x+5)}$, $x \neq -5, 2$ (b) $\frac{-1}{x(x^2-x+1)}$, $x \neq 0$
(c) $\frac{x-1}{x-4}$, $x \neq 2, 4$ (d) $\frac{(x-1)^2}{(x-2)^2}$, $x \neq -2, -1, 2$

Q3. (a) $\frac{-(2x+h)}{x^2(x+h)^2}$, $h \neq 0$ (b) $\frac{-2}{(2x+2h+3)(2x+3)}$, $h \neq 0$

Q4. (a) $\frac{3\sqrt{x}}{x}$ (b) $\frac{3-14\sqrt{x}+8x}{1-16x}$

Q5. (a)
$$\frac{1}{\sqrt{x+h+4}+\sqrt{x+4}}$$
(b)
$$\frac{1}{h[(x+h)^{\frac{2}{3}}+x^{\frac{1}{3}}(x+h)^{\frac{1}{3}}+x^{\frac{2}{3}}]}$$
(c)
$$-\frac{1}{x(x+h)}$$

3.7 Quadratic Equations

Q1. (a) $x = 0, 3$ (b) $x = \pm\frac{3}{2}$ (c) $x = -2, 5$ (d) $x = -11, -2$

Q2. (a) $x = -6, -2$ (b) $x = \frac{-5 \pm \sqrt{13}}{2}$ (c) $x = \frac{-2 \pm \sqrt{10}}{3}$

(d) $w = -10$ (twice)

Q3. (a) $x = 2, 5$ (b) $z = \pm\sqrt{27} - 4$ (c) $w = \dfrac{1}{4}$

Q4. (a) $4\sqrt{5}i$ (b) $10i$

Q5. (a) $10 - i\sqrt{53}$ (b) $\dfrac{1 - i\sqrt{3}}{4}$

Q6. (a) $x = \pm 2i$ (b) $x = \dfrac{-1 \pm i\sqrt{14}}{3}$ (c) $v = \dfrac{-1 \pm i\sqrt{5}}{4}$

3.8 Inequalities, Intervals and the Test-Point Method

Q1. (i) $x < 7$ (ii) $x \geq -1$ (iii) $-3 \leq x < 12$

Q2. (i) $-5 \leq x \leq 2$ (ii) $x^2 - 4x + 9 > 0 \quad \forall x \in R$
(iii) $-\sqrt{8} - 1 \leq x \leq \sqrt{8} - 1$
(iv) $x \in [-3, 0] \cup [3, \infty)$ (v) $x > 1$

Q3. (i) $-2 < x \leq 5$ (ii) $x \in (-3, 1] \cup (3, \infty)$ (iii) $x \in (-1, 2) \cup (4, \infty)$
(iv) $x \in [-4, 3) \cup (5, \infty)$

3.9 Absolute Value

Q1. (i) $x \leq -7$ (ii) $-1 < x < 5$ (iii) $x \in (-\infty, \frac{4}{3}] \cup [\frac{8}{5}, \infty)$
(iv) $x \in (-\infty, -2) \cup (6, \infty)$

4.1 Coordinate Systems

Q1. $\mid AB \mid = \sqrt{34}$, $\mid BC \mid = \sqrt{17}$, $\mid AC \mid = \sqrt{17}$,
hence
$\mid AB \mid^2 = \mid AC \mid^2 + \mid BC \mid^2$,
and
$\triangle ABC$ is right-angled.

Q2. $(x + 1)^2 + (y - 4)^2 = 25$

Q3. Circle center $(-3, -2)$, radius 2.

4.2 Equation of a Line

Q1. Slope is $2x + h$.

Q2. Slope is $-\dfrac{1}{x(x+h)}$.

Q3. (a) Unique intersection point at $\left(\dfrac{22}{5}, -\dfrac{13}{5}\right)$ (b) Lines parallel

4.3 Functions

Q1. (a) $f(0) = 1$, $f(x^2) = \dfrac{1}{1+x^2}$, $f(-2) = -1$, $f(\sqrt{x}) = \dfrac{1}{1+\sqrt{x}}$

(b) $h(0) = 1$, $h(-4) = 13$, $h(z^5) = 1 + z^5 + z^{10}$,
$h(1+u^2) = 1 + (1+u^2) + (1+u^2)^2$

(c) $g(0) = 1$, $g(n^3 - 1) = \sqrt{n^3}$, $g(\frac{1}{t}) = \sqrt{\frac{1}{t} + 1}$

Q2. (a) Domain is R. (b) Domain is $\{t \in R : t \neq \pm 1\}$.
(c) Domain is $\{u \in R : u \geq -1,\ u \neq 3\}$

Q3. $\dfrac{1}{\sqrt{x+h} + \sqrt{x}}$

Q4. Domain is $\{x \in R : x \neq 0\}$, range consists of the two elements ± 1

Q5. (a) even (b) neither even nor odd (c) odd

4.4 Operations with Functions

Q1. (a) $(f+g)(x) = -2x - 5$, domain is R;

$(f \cdot g)(x) = -4x(2x-5)$, domain is R;

$(\frac{f}{g})(x) = \dfrac{2x-5}{-4x}$, domain is $R \backslash \{0\}$.

(b) $(f+g)(x) = \dfrac{x}{x+1} + \dfrac{x-1}{x}$, domain is $R \backslash \{-1, 0\}$;

$(f \cdot g)(x) = \dfrac{x}{x+1} \cdot \dfrac{x-1}{x}$,domain is $R \backslash \{-1, 0\}$;

$(\frac{f}{g})(x) = \dfrac{x^2}{(x-1)(x+1)}$, domain is $R \backslash \{-1, 0, 1\}$.

Q2. (a) $(f \circ g)(x) = \dfrac{1}{2x}$, domain is $R \backslash \{0\}$,

$(g \circ f)(x) = \dfrac{1}{2x}$, domain is $R \backslash \{0\}$

(b) $(f \circ g)(x) = \sqrt{x^4 + 1}$, domain is $\{x \in R : x \geq -1\}$,

$$(g \circ f)(x) = (x+1)^2, \text{ domain is } \{x \in R : x \geq -1\}.$$

(c) $(f \circ g)(x) = \dfrac{x-1}{3x-1}$, domain is $\{x \in R : x \neq 0, \frac{1}{3}\}$,

$(g \circ f)(x) = -\dfrac{2}{x}$, domain is $\{x \in R : x \neq 0, -2\}$.

Q3. $(f \circ g)(x) = x = (g \circ f)(x)$, domain is R in both cases. Hence,

$$g \circ f = f \circ g$$

5.1 Graphs of Polynomial Functions

Q1. (a) $x = 1, 2, 3$ (b) $x = 0, 1$(twice) (c) $x = -4, -2, 0$
(d) $x = 0, 9$

Q2. (a) $p(1) = -4 < 0$, $p(2) = 13 > 0$ so $p(x)$ has a zero in $(1, 2)$
(b) $p(1) = -6 < 0$, $p(2) = 78 > 0$ so $p(x)$ has a zero in $(1, 2)$

5.2 Rational Functions and Division of Polynomials

Q1. (a) $\underbrace{1}_{\text{quotient}} + \underbrace{\dfrac{1}{x+1}}_{\text{remainder}}$ (b) $\underbrace{2x+2}_{\text{quotient}} + \underbrace{\dfrac{7}{2x-3}}_{\text{remainder}}$

(c) $x+1$ (remainder is zero) (d) $x-1$ (remainder is zero)

5.3 Factoring Higher Order Polynomials

Q1. (a) $p(1) = 0$; $(x-1)(2x^2 - 5x + 4)$ (b) $p(3) = 0$; $(x-3)(x-2)^2$

Q2. (a) $(x-3)(x-2)$ (b) $2(x-3)(x-4)$ (c) $(x-2)(x+2)(x-1)(x+1)$
(d) $(x-3)(x+3)(x-2)(x+2)$

Q3. (a) $p(1) = 0$, so $x-1$ is a factor. Divide to obtain

$$p(x) = (x-1)(x-3)(x+1).$$

Hence, solutions (zeros) of $p(x) = 0$ are $x = -1, 1, 3$.

(b) $p(\frac{1}{2}) = 0$, so $x - \frac{1}{2}$ is a factor. Divide to obtain

$$p(x) = (x - \frac{1}{2})2(x^2 + 2x - 1).$$

Hence, solutions (zeros) of $p(x) = 0$ are $x = \frac{1}{2}, -1 \pm \sqrt{2}$ (using quadratic formula).

(c) $p(\frac{1}{2}) = 0$, so $x - \frac{1}{2}$ is a factor. Divide to obtain

$$p(x) = (x - \frac{1}{2})2(x^2 - x - 1).$$

Hence, solutions (zeros) of $p(x) = 0$ are $x = \frac{1}{2}, \frac{1 \pm \sqrt{5}}{2}$ (using quadratic formula).

6.2 Trigonometric Identities

Q2. Let $x = \frac{7\pi}{6}$, $\sin(\frac{7\pi}{6}) = -\frac{1}{2}$; $\quad \sqrt{\sin^2(\frac{7\pi}{6})} = \sqrt{\frac{1}{4}} = \frac{1}{2}$.
The formula is true only for $\sin x \geq 0$ (recall $\sqrt{x^2} = |x|$ - see §3.9).

7.2 Logarithmic Functions

Q1. (a) 2 (b) x^{10} (c) x (d) $2\ln(x+2)$, $x > -2$ (e) $3\ln x$, $x > 0$

Q2. Take logarithms of both sides and solve for x to obtain $x = 2\left[\frac{\ln 2}{\ln(\frac{2}{3})}\right]$

Q3. Domain is $x \in (0, 4)$ (from the fact that we require $\frac{x}{4-x} > 0$).

Answers to Assessment Test

1. (d) 2. (b) 3. (d) 4. (b) 5. (b) 6. (a) 7. (c)

8. (a) 9. (d) 10. (d) 11. (a) 12. (a) 13. (c)

14. (c) 15. (b) 16. (a) 17. (b) 18. (d) 19. (c)

20. (d) 21. (c) 22. (b) 23. (a) 24. (c) 25. (d)

9. INDEX OF METHODS AND FORMULAE

/cont.

PART 2

MIDTERM AND FINAL PRACTICE EXAMINATIONS WITH FULLY WORKED SOLUTIONS

MAKING THE MOST OF THE PRACTICE EXAMS

The examinations cover the full range of topics offered in a regular beginning calculus course. These include

Midterm Exam Topics:

Functions
Inequalities
Limits
(Continuity)
Differentiation
  – *from limit definition, product and quotient rules, chain rule*
  – *implicit differentiation*
Related Rates
Differentials

Final Exam Topics:

Selected Midterm Topics (usually those in italics)
Rolle's Theorem
Mean Value Theorem
Intermediate Value Theorem
Tangent Lines
Maxima and Minima
Graphing
  – *domain, range, local extrema, points of inflection, asymptotes, sketch*
Definition of the integral
Antiderivatives
The Fundamental Theorem of Calculus
Definite Integrals
Areas between curves

(*Italics* indicate frequently examined topics.)

Early Transcendentals Option

Students taking beginning calculus with the early transcendentals option may cover **in addition** to the above: polar coordinates, Taylor polynomials, l'Hôpital's rule and, of course, transcendental functions (e.g. logarithm, exponential and inverse trigonometric functions) together with their use in limits, differentiation, integration and curve sketching. These additional topics (except possibly 'polar coordinates') are usually final exam topics. Midterm exams 3 and 4 and final exams 2 and 5 include options designed to accommodate this choice of presentation (although students in this category should not restrict themselves to these exams alone – the other examinations are just as important for practice and development of the relevant techniques which form the basis of beginning calculus).

The problems presented in the examinations are representative of the material in most beginning calculus courses. It should be noted, however, that an individual instructor might emphasise certain topics more than others (and may include some in addition) in any particular midterm/final exam. For this reason, students should stay in touch with the individual instructor's requirements.

Working With the Examinations

(a) The examinations are rated on a scale of 1 to 5. Levels 3 and 4 represent the typical standard while level 5 is more challenging. Start with levels 3 and 4 and move onto level 5 when you have gained the necessary confidence.

(b) All examinations include an allotted time. This will give you some idea of the rate at which you should be completing the problems. Don't worry too much about this, however. The point here is to expose yourself to as many typical examination questions (and their solutions!) as possible, i.e. this is a learning exercise. With this in mind, try to struggle a little before consulting the solutions. Later, when you have sufficient confidence, there is nothing to stop you from doing any of the examinations under 'actual examination conditions' – you can even grade them yourself!

(c) Calculators are not necessary in any of the problems.

(d) Students taking the early transcendentals option should pay particular attention (but not restrict themselves) to midterm exams 3 and 4 and final exams 2 and 5 which offer, in addition, optional questions designed to accommodate this choice of presentation (note that the midterms are almost unaffected by the early transcendentals option). Students taking the 'regular' beginning calculus route (i.e. without transcendentals) should merely avoid any of the options offered in these particular exams.

(e) Students who have two midterms can pick and choose questions on which to practice given the individual instructor's requirements.

(f) At final examination time don't forget to use selected midterm examination questions for practice. Remember, the final examination usually includes questions examining topics offered before the midterm examination.

Using the Solutions

(a) The solutions are written more to provide teaching assistance than to furnish a set of answers. Many of the solutions are detailed and include the relevant theory/formulae used in arriving at the correct answer. The reason for this comes from the author's own experience as an instructor – students learn most when actually applying the theory to examples **particularly** when they have either someone to ask or a full set of 'teaching solutions'. Study the solutions carefully. Try to mimic the steps and the reasoning behind each solution. Practice in this context will develop a clear and logical approach to each **type** or **class** of problem.

(b) The solutions to the first examination in each set are slightly more detailed than those corresponding to the subsequent four examinations. This is mainly to avoid repetition of formulae/techniques explained 'first time around'. Students can always consult the solutions to the first examination and/or Part 1 of the text when requiring more detail on a particular technique.

(c) Whenever they appear, C or c represent arbitrary constants of integration.

(d) Remember, this publication is not intended to replace but rather to complement the regular offerings of a beginning calculus course. If you require more explanation of the theory, consult your instructor or the recommended textbook.

Strategies for Success

Clearly, what you do during term-time will significantly affect the outcome of your examination. There are several things that you can do, but in mathematics three things stand out above all others. They are:

(i) *Solve as many problems as you can* – i.e. assignment questions, questions from the textbook, etc. The more problems you solve, the more you will become fluent in the relevant techniques (learning mathematics is very much like learning a new language). This will make your exam-preparation easier.

(ii) *Ask!* Don't be afraid to ask questions. Your instructor is there to answer your questions!

(iii) *Work through practice/sample examinations!* This will 'fine-tune' your exam preparation, serve to sharpen existing skills and act as a performance indicator.

MIDTERM EXAMINATION #1

Time: 50 minutes **Level of difficulty**: 3

Value

(15%) 1. Find (a) $\lim_{x \to 2} \dfrac{\frac{1}{x^3} - \frac{1}{8}}{x - 2}$

(b) $\lim_{x \to 1^-} \dfrac{|x-1|}{x^2 - 5x + 4}$

(c) the domain of the function:

$$f(x) = \frac{\sqrt{6 - x^2}}{x - 2}$$

(20%) 2. (a) Find $f'(x)$ if

(i) $f(x) = x^3(1 - x^3)^{\frac{4}{3}}$

(ii) $f(x) = \dfrac{(x^3 + 1)}{\csc x}$

(iii) $f(x) = \sqrt{\sin(\cos^2 x)}$

(b) Find the equation of the line tangent to the graph of

$$x^2y^2 = (y + 1)^2(9 - y^2)$$

at the point $(0, -3)$.

(12%) 3. Discuss the continuity of

$$f(x) = \begin{cases} x^2, & x < 0, \\ 2x - 1, & 0 \le x \le 2, \\ 2x^2 - 5, & x > 2. \end{cases}$$

(12%) 4. Ship A is 10 miles north of ship B at 1300 hrs. Ship A sails east at 18 mph while ship B sails north at 12 mph. How fast is the distance between them changing at 1500 hrs?

Value

(16%) 5. (a) Use differentials to approximation $\sqrt{10}$.

(b) Show that Newton's method fails when applied to the equation $\sqrt[4]{x} = 0$ with any initial approximation $x_1 \neq 0$.

(25%) 6. (a) Use the *limit definition* of the derivative to find $f'(x)$ if $f(x) = \sqrt{2x+3}$.

(b) What is the equation of the tangent line to the graph of $y = \sqrt{2x+3}$ at the point $(3, 3)$?

(c) What is $\dfrac{d^2y}{dx^2}$ if $\sqrt{x} + \sqrt{y} = 321$?

MIDTERM EXAMINATION #2

Time: 50 minutes **Level of difficulty**: 4

Value

(18%) 1. Find each of the following limits:

(a) $\lim\limits_{x \to 2} \dfrac{\sqrt{x+2}-2}{x-2}$

(b) $\lim\limits_{x \to 1} \dfrac{|x-1|}{x^2-1}$

(c) $\lim\limits_{x \to 0} \dfrac{\cos^2 4x - 1}{4\cos^2 5x - 4}$

(20%) 2. (a) Use the *limit definition* of the derivative to find $f'(x)$ if $f(x) = \dfrac{1}{3x+2}$.

(b) Find the equation of the tangent line to the curve $y = \dfrac{1}{3x+2}$ at the point $\left(-\dfrac{1}{3}, 1\right)$.

(24%) 3. In each of the following cases find $\dfrac{dy}{dx}$.

(a) $y = \dfrac{x^3+1}{x^2-1}$

(b) $y = (2x^3+1)(3x+1)^{\frac{1}{3}}$

(c) $y = \sqrt{1+\sin^2 5x}$

(d) $x^2 \tan y = x^3 + y^2$

(20%) 4. (a) Define precisely what it means to say that a function $f(x)$ is continuous at $x = a$.

(b) Suppose

$$f(x) = \begin{cases} -x+1, & x < -1, \\ x^2+1, & -1 < x \le 0, \\ x^2+2, & 0 < x \le 1, \\ 2, & x > 1, \end{cases}$$

(i) Sketch the graph of $y = f(x)$.

Value

(ii) Indicate where $f(x)$ is not continuous – give reasons for your answers.

(iii) Is $f(x)$ continuous on $[0, 1]$? Explain.

(18%) 5. A water container has the shape of an inverted cone. Water is leaking out at a rate of $\frac{1}{2}$ m^3 per hour. The container has a height of 6 metres and the radius at the top of the container is 1 metre. How fast is the water level falling when the water in the container is 4 metres deep?

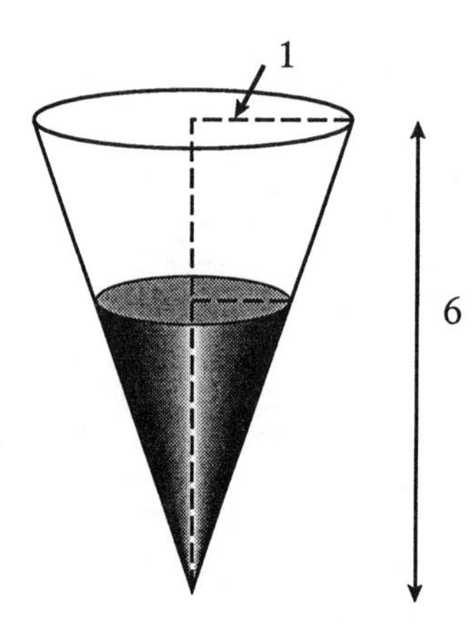

MIDTERM EXAMINATION #3

Time: 90 minutes **Level of difficulty**: 4

Value

(17%) 1. Find the following derivatives. You need not simplify your answer.

(a) If $y = \dfrac{\sin 2x}{\sqrt{x + \cos 3x}}$, find y'.

(b) If $y = \tan\left(\dfrac{1}{x^2}\right)$, find y', y''.

(c) If $x = \sin[\theta \tan\theta]^2$, find $\dfrac{dx}{d\theta}$.

(17%) 2. Evaluate the following limits. Show your work.

(a) $\lim_{x \to 0} \dfrac{1}{x}(\sqrt{2+x} - \sqrt{2-x})$

(b) $\lim_{x \to 0} (x^2 + 3x) \cot 6x$

(c) $\lim_{h \to 0} \dfrac{(1+h)^{-2} - 1^{-2}}{h}$

(11%) 3. Let the function f be defined on $\left(-\frac{\pi}{2}, \frac{\pi}{2}\right)$ by

$$f(x) = \begin{cases} \dfrac{\sin 6x}{\sin x}, & x \neq 0, \\ c, & x = 0, \end{cases}$$

where c is some constant. Find c such that f is a continuous function on $\left(-\frac{\pi}{2}, \frac{\pi}{2}\right)$. Justify your conclusion.

(11%) 4. Find $f'(x)$ using the *limit definition* of a derivative if $f(x) = \dfrac{1}{\sqrt[3]{x}}$.

(17%) 5. (a) Determine all points where the graph of the equation

$$\sin x \cos y + (1 - x^2)y^2 = \pi^2$$

intersects the y-axis.

(b) Find the equation of the tangent line to the graph at each of the points found in (a).

Value

(11%) 6. (a) At a certain instant, the surface area S of a spherical balloon is increasing at the same rate as its radius r is increasing. What is the radius at that instant?

(b) A spherical raindrop accumulates moisture (through condensation) at a rate proportional to its surface area S. Prove that the radius r increases at a constant rate.

(16%) 7. (a) The kinetic energy K of a mass m moving with speed v is given by

$$K = \frac{1}{2}mv^2.$$

Use differentials to estimate the percentage increase in kinetic energy of a mass m if its speed is decreased from 160 km/h to 157 km/h.

(b) Use differentials to estimate $\sin(58°)$.

OR*

Sketch the curve with polar equation $r = 2\cos\theta + 2\sin\theta$.

* For students taking the early transcendentals option.

MIDTERM EXAMINATION #4

Time: 90 minutes **Level of difficulty**: 5

Value

(14%) 1. (a) Let $f(x) = \dfrac{x}{x+2}$ and $g(x) = \dfrac{\sqrt{x-1}}{x}$. Find $(f \circ g)(x)$ and its domain.

(b) If $f(x) = \cos\left[1 + \sqrt{x}\sin(x^3+4)\right]$, find $f'(x)$. (You need not simplify your answer.)

(14%) 2. (a) Show that the sum of the x- and y-intercepts of any tangent line to the curve $\sqrt{x} - \sqrt{y} = a$, is equal to a^2.

(b) Find $\dfrac{d^2y}{dx^2}$ if $y = \sqrt[3]{1+\cos x}$. (You need not simplify your answer.)

(14%) 3. Find $f'(x)$ using the *limit definition* of the derivative if

$$f(x) = \frac{1}{\sqrt{x-a}} \qquad (a \text{ is a constant}).$$

State the domain of the function and of its derivative.

(16%) 4. Evaluate each limit or explain why there is no limit.

(a) $\displaystyle\lim_{x\to 0} \frac{2-\sqrt{1-x^2}}{x}$

(b) $\displaystyle\lim_{x\to 0} \frac{\tan^2(8x)}{3x}$

(c) $\displaystyle\lim_{x\to 1^-} \left[\frac{(x^2-2x+1)^{-1} - (x-1)^{-1}}{(x^2-9)}\right]$

(14%) 5. Consider the function

$$f(x) = \begin{cases} x^2 - x + 1, & x < 1, \\ 1, & x \geq 1. \end{cases}$$

(a) Is $f(x)$ continuous at $x = 1$? Why?

(b) Is $f(x)$ differentiable at $x = 1$? Why?

Value

(14%) 6. A right circular cone of radius 2m and height 4m has its tip oriented downward and is filled with oil. The oil is escaping at a rate of $5\text{cm}^3/\text{sec}$. How fast is the area of the top surface of the oil changing when the height of oil in the cone is 600cm?

(14%) 7. (a) If $f(x) = |x^3 - x^2|$, find f' and f''. What are their domains?

(b) Does the equation

$$x^8 + 3x^2 - x - 2 = 0$$

have any real roots?

OR*

Sketch the curve with equation $(x^2 + y^2)^6 = 16x^4y^4$.

* For students taking the early transcendentals option.

MIDTERM EXAMINATION #5

Time: 50 minutes **Level of difficulty**: 3

Value

(25%) 1. Find the limit if it exists. Explain why there is no limit otherwise.

(a) $\lim_{x\to 4} \dfrac{\sqrt{x}}{2+2x-x^2}$

(b) $\lim_{x\to -5} \dfrac{1}{x+5}$

(c) $\lim_{x\to 0} \dfrac{|x|}{2x}$

(d) $\lim_{x\to 0} \dfrac{x^2+\sin x}{\sqrt{x+4}-2}$

(25%) 2. (a) Is the following function continuous at $x = 3$? Give reasons for your answer.

$$f(x) = \begin{cases} \dfrac{x^2-x-6}{x-3}, & x \neq 3, \\ 3, & x = 3. \end{cases}$$

Sketch the graph of the function.

(b) Is the following function continuous at $x = -2$? Give reasons for your answer.

$$f(x) = \frac{x^2-4}{x+2}.$$

(30%) 3. In each of the following, find $\dfrac{dy}{dx}$.

(a) $y = \left(x + \dfrac{1}{x^2}\right)^{\sqrt{8}}$

(b) $y = \sqrt{x\sqrt{x\sqrt{x}}}$

(c) $y = \dfrac{\sin x}{\sin(x - \sin x)}$

(d) $x \tan y = y - 1.$

Value

(20%) 4. (a) Find y'' if $x^6 + y^6 = 1$.

(b) Find the equation of the tangent to the curve defined by

$$\cos x + \cos y = x^2 + 1$$

at the point $\left(0, \frac{\pi}{2}\right)$.

FINAL EXAMINATION #1

Time: 2 hours **Level of difficulty**: 4

Value

(10%) 1. (i) Evaluate the following limits.

(a) $\lim_{x \to \infty} \left(\sqrt{x^2 + 3x - 1} - x\right)$

(b) $\lim_{x \to -\infty} \left(\sqrt{x^2 + 3x - 1} - x\right)$

(c) $\lim_{h \to 0} \dfrac{\sin \pi h^2 \tan 3h^2}{h^3}$

(ii) Let

$$f(x) = \begin{cases} 1 - \sin x, & -2\pi \le x < 0, \\ 1, & x = 0, \\ x^2 - 1, & 0 < x \le 2. \end{cases}$$

(a) Is $f(x)$ continuous at $x = 0$? Explain.

(b) Is $f(x)$ differentiable at $x = 0$? Explain.

(15%) 2. Evaluate the following definite integrals:

(a) $\displaystyle\int_0^1 |x^3 - x - 3x^2 + 3| dx$

(b) $\displaystyle\int_0^1 \frac{2x^3 + 9x^2 + 5}{(x^4 + 6x^3 + 10x + 7)^{\frac{1}{2}}} dx$

(c) $\displaystyle\int_{-1}^1 x^{-\frac{8}{9}} \sin\left(x^{\frac{1}{9}}\right) dx$

(d) $\displaystyle\int_{-a}^a x^3 \cos(x^3) dx, \quad a \in \mathbb{R}.$

(15%) 3. Find the following.

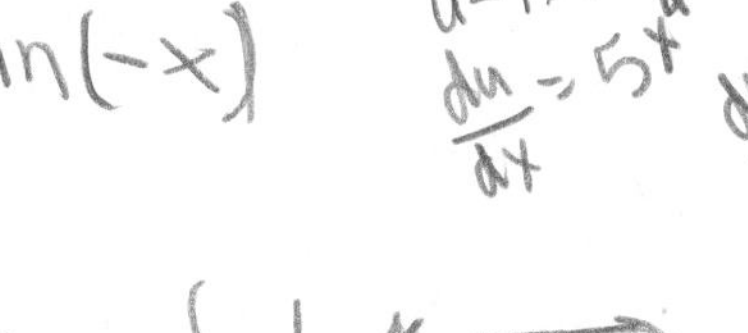

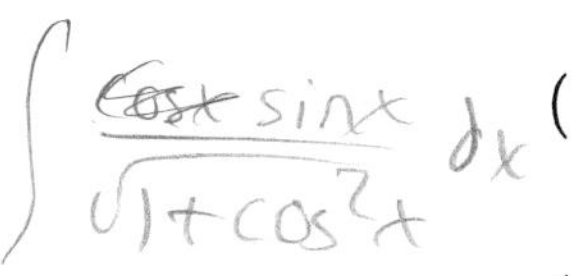

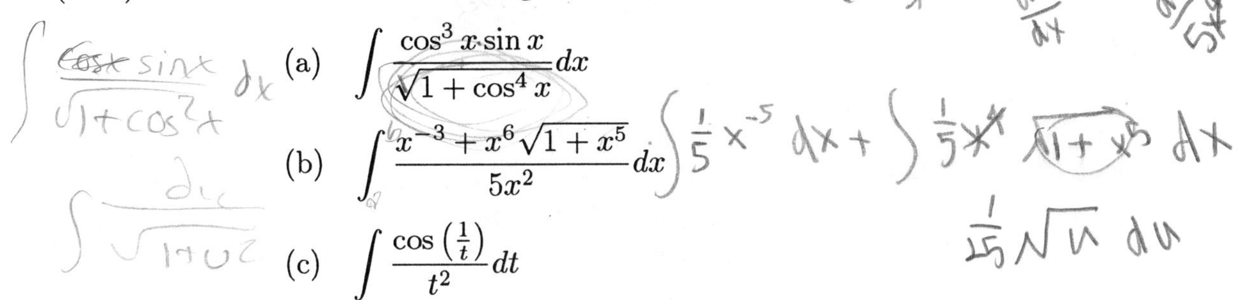

(a) $\displaystyle\int \frac{\cos^3 x \sin x}{\sqrt{1 + \cos^4 x}} dx$

(b) $\displaystyle\int \frac{x^{-3} + x^6\sqrt{1 + x^5}}{5x^2} dx$

(c) $\displaystyle\int \frac{\cos\left(\frac{1}{t}\right)}{t^2} dt$

Value

(d) $\int \sin^4 t \cos^3 t dt$

(e) $f'(x)$ where $f(x) = \int_x^{x^2+1} \tan\left(\sqrt{u^2+1}\right) du$

(20%) 4. Discuss the graph of $y = f(x) = \dfrac{7x^2}{x^2-1}$ under the following headings.

(a) Domain of $f(x)$.

(b) x and y-axis intercepts.

(c) Asymptotes.

(d) Intervals of increase or decrease.

(e) Local maximum and minimum values.

(f) Concavity and points of inflection.

Sketch the curve.

(10%) 5. An open-top cylindrical can is made to hold b litres of oil. Find the dimensions that will minimize the cost of the metal.

(10%) 6. Find the limit if it exists. If the limit does not exist, explain why.

(a) $\lim_{\theta \to 0} \left(\theta^2 \cot \theta + 5\right)$

(b) $\lim_{x \to 0} \left(\dfrac{1 - \sin x}{x^3}\right)$

(c) $\lim_{x \to 1^+} \dfrac{1 - \sqrt{x}}{1 - x}$

(d) $\lim_{x \to -\infty} \dfrac{x - 1}{x^2}$

(10%) 7. Find the area bounded by the graphs of $y = x^3$ and $y = x$.

(10%) 8. The height h of a triangle is increasing at a rate of 5cm/min while the area of the triangle is increasing at a rate of 8cm^2/min. At what rate is the base of the triangle changing when the height is 50cm and the area 500cm^2?

FINAL EXAMINATION #2

Time: 2 hours **Level of difficulty**: 4

Value

(30%) 1. Evaluate when possible:

(a) $\lim\limits_{x\to 0} \dfrac{x - x\cos 3x}{x + x\cos 3x}$ **OR*** $\lim\limits_{x\to 0} \dfrac{\sin x - x - x^3}{x^2}$

(b) $\lim\limits_{x\to 2} \left(\dfrac{1}{x-2} - \dfrac{2}{x^2-4}\right)$ **OR*** $\lim\limits_{x\to\infty} \dfrac{(\ln x)^2}{x}$

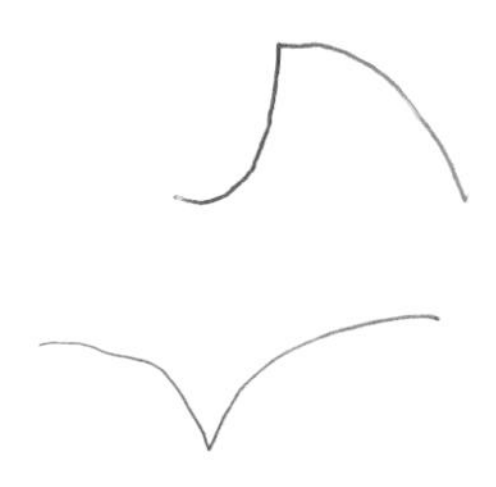

(c) $\lim\limits_{x\to -\infty} \dfrac{8x^3 + 2x^2 - 1}{(x^9 - 4x + 11)^{\frac{1}{3}}}$ **OR*** $\lim\limits_{x\to 0^+} (3x+1)^{\frac{2}{x}}$

(d) $\displaystyle\int_0^2 |x^3 - 3x^2 + 2x|dx$ **OR*** $\displaystyle\int_1^8 e^{\sqrt[3]{x}} x^{-\frac{2}{3}} dx$

(e) $\displaystyle\lim_{n\to\infty} \sum_{i=1}^{n} \frac{8}{n}\left[3\left(2 + \frac{2i}{n}\right)^2 + 2\left(2 + \frac{2i}{n}\right)\right].$

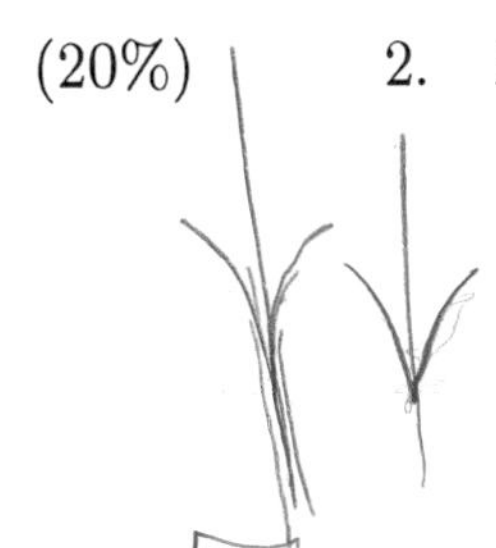

(20%) 2. Find: (a) $f'(x)$ if $f(x) = \left[10 + \left(1 + \sqrt{3+x}\right)^{\frac{3}{2}}\right]^{\frac{3}{2}}$
OR* $f(x) = e^{\sqrt{x}}\cos(x^2)$

(b) $g'(x)$ if $g(x) = \cos\left[\tan\left(\sin\left(x^2+1\right)\right)\right]^{\frac{1}{2}}$
OR* $g(x) = \ln[\ln(x^4)]$

(c) $\dfrac{dy}{dx}$ if $x^2y = \cos(xy)$
OR* $e^{x\sin y} + x^3y^2 = \ln(x^2+y^2)$

(d) $y'(x)$ if $y(x) = \displaystyle\int_{\sqrt{x}}^{1} \sqrt{2 + \tan^2(t^2)}dt$

(Do not simplify your answers.)

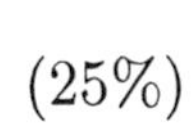

(25%) 3. Find: (a) $\displaystyle\int x^3\sin(3x^4+10)dx$ **OR*** $\displaystyle\int_{e^2}^{e^3} \frac{dx}{x\ln x}$

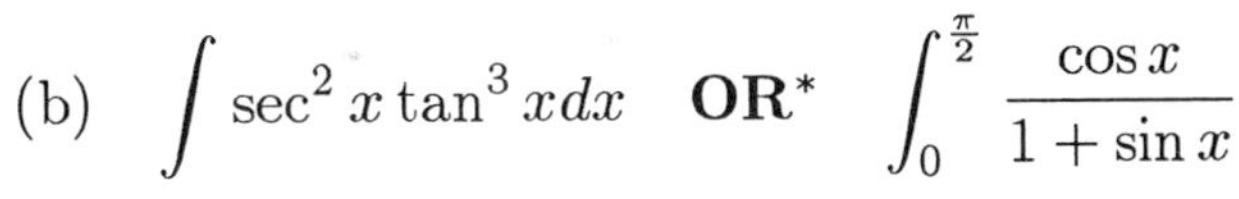

(b) $\displaystyle\int \sec^2 x\tan^3 x dx$ **OR*** $\displaystyle\int_0^{\frac{\pi}{2}} \frac{\cos x}{1+\sin x}$

(c) $\displaystyle\int_{-\pi}^{\pi} \frac{t^4\sin t}{1+t^8}dt$ **OR*** $\displaystyle\int_0^1 \frac{e^{\sin^{-1}x}}{\sqrt{1-x^2}}dx$

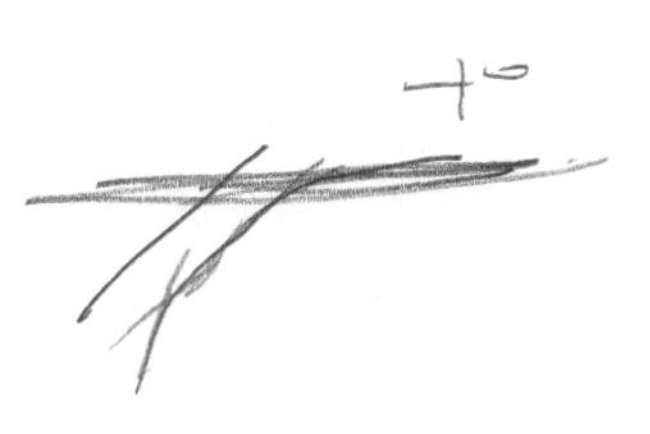

Value

(d) $\displaystyle\int \frac{\sqrt{1+x^{-5}}}{x^6}dx$ **OR*** $\displaystyle\int \frac{e^x}{e^x+2}dx$

(e) $\displaystyle\int \sin x\,[\sin(\cos x)]\,dx$ **OR*** $\displaystyle\int \frac{\sqrt{\arctan x}}{1+x^2}dx$

(10%) 4. Sketch the graph of $y = f(x) = \dfrac{x}{x-1}$ **OR***
$y = f(x) = \ln(\cos x)$ identifying any critical points, asymptotes, x and y-intercepts, points of inflection and local maxima and minima.

(8%) 5. Find the total area bounded by the curves $y = x - 2$ and $y^2 - 2x - 4 = 0$.

OR*

Find the Taylor polynomial of degree three, $T_3(x)$, for the function $f(x) = e^x \sin x$ about the point $c = 0$.

(7%) 6. Find the ratio of the height (h) to the radius (r) of the minimum surface-area cone of constant volume b.

(Hint: Surface area S of cone is given by $S^2 = \pi^2(r^4 + r^2h^2)$ and $b = \frac{1}{3}\pi r^2 h$.)

* For students taking early transcendentals – although the transcendentals options are important, you should practice with **both** versions of this exam.

FINAL EXAMINATION #3

Time: 2 hours **Level of difficulty**: 4

Value

(15%) 1. (a) If $f(x) = \frac{1}{8x}$, find $f^{(n)}(x)$, where n is a positive integer.

(b) Consider the function $f(x) = \int_3^{\sqrt{x}} 2t \cos t dt$, $x \in [0, \pi]$.

(i) For what values of x does this function have local extrema?

(ii) Are there any points of inflection?

(30%) 2. Find: (a) $\int \sin^3 x dx$

(b) $\int \left[\sqrt{\tan\theta} + (\tan\theta)^{\frac{5}{2}}\right] d\theta$

(c) $\int_0^1 \frac{x^{\frac{1}{2}}}{\sqrt{1 + x^{\frac{1}{2}}}} dx$

(d) $\int x^7 \sqrt{1 + x^2} dx$

(e) $\int \frac{\cos^3 x \sin x}{[1 + \cos^4 x]^{\frac{3}{2}}} dx$

(f) $\int_0^{\pi} \sqrt{\cos^2 x} dx$

(g) $\lim_{x \to \infty} \frac{2 - \sqrt{x}}{2 + \sqrt{x}}$

(h) $\lim_{\theta \to \infty} 3\theta \sin\left(\frac{1}{\theta}\right)$

(20%) 3. Sketch the graph of

$$y = f(x) = x^{\frac{4}{3}} - 4x^{\frac{1}{3}}$$

indicating where the function is increasing, decreasing, concave up, concave down, has local maxima, local minima, inflection points and asymptotes.

Value

(8%) 4. Show that the equation

$$3x^5 + 18x - 1 = 0$$

has *exactly* one real root.

(20%) 5. Evaluate the following (if they exist) or explain why they don't exist.

(a) $\lim_{x \to 1} \frac{x^{\frac{1}{3}} - 1}{(x - 1)}$

(b) $\lim_{x \to -\infty} \left(\sqrt{2x^2 - 3x + 1} + x\sqrt{2}\right)$

(c) $\lim_{h \to 0} \frac{\sin\left(\frac{\pi}{2} + h\right) - 1}{h}$

(d) $\lim_{x \to -\infty} \frac{3x^3}{\sqrt[3]{x^9 + 2x + 3}}$

(7%) 6. (a) When a sample of gas is compressed at a constant temperature, the pressure $P(t)$ and the volume $V(t)$, at time t, are related by the equation

$$PV = C, \quad C \text{ is constant.}$$

Suppose that at a certain instant $V = 10\text{cm}^3$ and $P = 8$ Pascals with the latter falling at a rate of 2 Pascals/minute. At what rate is the volume increasing at this instant?

(b) Verify the Mean Value Theorem for the function

$$f(x) = \cos x \quad \text{on} \quad \left[-\frac{\pi}{2}, 0\right].$$

FINAL EXAMINATION #4

Time: 2 hours **Level of difficulty**: 5

Value

(10%) 1. Determine all line asymptotes of each of the following functions and sketch the graph. (Note: no derivatives are required.)

(i) $f(x) = \dfrac{6x}{x-2}$

(ii) $g(x) = \dfrac{x^2-2}{x+1}$

(15%) 2. Suppose a spherical ball loses air in such a way that its volume decreases at a rate proportional to its surface area. If it takes the ball 4 hours to deflate to half its original volume, how much longer will it take for the ball to deflate completely?

(15%) 3. Find the following limits or explain why they don't exist.

(a) $\displaystyle\lim_{x\to 0} \frac{2x}{|x-2|-|x+2|}$

(b) $\displaystyle\lim_{x\to -2} \left(\frac{1}{x+2} - \frac{23}{(x^2-4)}\right)$

(c) $\displaystyle\lim_{x\to a} \frac{f(x)-f(a)}{\sqrt[3]{x}-\sqrt[3]{a}}, \quad a > 0$

(30%) 4. Find the following.

(a) $\displaystyle\int \frac{3x^6 - 2x^{\frac{1}{2}}}{x^4} dx$

(b) $\displaystyle\int \frac{x+\sqrt{x^2+2}}{\sqrt{x^2+2}} dx$

(c) $\displaystyle\int \cos x \sec^2(\sin x) dx$

(d) $\displaystyle\int_{-\frac{\pi}{3}}^{\frac{\pi}{3}} \sin^7 \theta d\theta$

(e) $f''(x)$ if $f(x) = \displaystyle\int_0^{\sqrt{x}} \left[\int_{\frac{1}{2}}^{\cos t} (1+u^2)^{\frac{3}{2}} du\right] dt$

(f) $\displaystyle\int_0^2 |x^2 - 3x + 2| dx$

Value

(10%) 5. Suppose a piece of wire of length 2m is cut into two pieces one of which is bent into an equilateral triangle and the other into a circle. How should the wire be cut so that the sum of the areas is a

(i) maximum?

(ii) minimum?

(10%) 6. (a) Find y' if $x^2y^3 + \cos\sqrt{xy} = \sqrt{x\sqrt{x}}$. (Do not simplify your answer.)

(b) Does continuity of a function at a point imply differentiability at that point? Illustrate with an example.

(10%) 7. Find the total area bounded by the curves $y = x$ and $y = x^{11}$.

FINAL EXAMINATION #5

Time: 2 hours **Level of difficulty**: 4

Value

(25%) 1. Find

(a) $\lim_{x\to 0+} \dfrac{\sqrt{1-\cos 2x}}{x \sec x}$ **OR*** $\lim_{x\to 0} \dfrac{3^x - 1}{x}$

(b) $\lim_{x\to\infty} \dfrac{2-\sqrt{x}}{2+\sqrt{x}}$ **OR*** $\lim_{x\to\infty} e^{-x}\ln x$

(c) $\lim_{x\to\infty} \sin\left(\dfrac{1}{x}\right)$ **OR*** $\lim_{x\to-\infty} \dfrac{e^{3x}-e^{-3x}}{e^{3x}+e^{-3x}}$

(d) $\lim_{x\to 3+} \left|\dfrac{x-3}{x+1}\right|$

(e) $f^{(n)}(x)$, $n \geq 1$, if $f(x) = \dfrac{x+1}{x}$

(20%) 2. Let $f(x) = \dfrac{1}{x-2} - x$ **OR*** $f(x) = xe^{x^2}$.

(a) What is the domain of $f(x)$?

(b) Write down the equations of all line asymptotes associated with the curve $y = f(x)$.

(c) Where is the graph of f increasing and where decreasing?

(d) Where is the graph of f concave up and where concave down?

(e) Locate any local extrema and any inflection points.

(f) Sketch the curve represented by $y = f(x)$.

(25%) 3. Find the following:

(a) $\displaystyle\int \frac{(\sqrt{\sin x} + \sin^2 x)}{\sqrt{1-\cos 2x}} \cos x\,dx$, $\sin x > 0$

OR* $\displaystyle\int \frac{\sin(\ln x^2)}{x} dx$

(b) $\displaystyle\int \sin^3 t \cos^2 t\,dt$ **OR*** $\displaystyle\int \frac{x}{x+1} dx$

(c) $\displaystyle\frac{d}{du}\int_u^{3u} \cos\sqrt{1+x^2}\,dx$ **OR*** $\displaystyle\frac{d}{dt}\int_1^{\ln t} \sin\sqrt{u^2+e^u}\,du$

Value

(d) $\displaystyle\int_0^1 \frac{3y^4}{(6+y^5)^8}dy$ **OR*** $\displaystyle\int e^x \sin^2(1+e^x)dx$

(e) $\displaystyle\int_{-1}^2 \sqrt{x^2}dx$ **OR*** $\displaystyle\int \frac{2+e^{-x}}{e^x}dx$

(10%) 4. A ladder of height h metres rests against a vertical wall. The bottom of the ladder is then pulled along the ground away from the wall at a constant rate of b metres/second. Derive an expression for the speed of the top of the ladder as it slides down the wall in terms of $\theta(t)$ and $\dfrac{d\theta}{dt}$ where $\theta(t)$ is the angle the ladder makes with the perpendicular to the ground at time t.

(10%) 5. (a) Consider the ellipse $x^2 + 3y^2 = 1$. Find the equations of the tangent lines with unit slope.

(b) Show that the equation

$$x^{91} + x^{31} + 8x - 5 = 0$$

has exactly one real root.

OR*

Find the Taylor polynomial $T_3(x)$ at $c = 1$ for the function $f(x) = \dfrac{1}{x}$ then sketch the graph of f against T_3.

(10%) 6. Use the Mean Value Theorem to show that if $f'(x) = 0\ \forall\ x \in (a, b)$, then f is constant on (a, b).

OR*

Differentiate with respect to θ:
(i) $g(\theta) = [\arctan\sqrt{\theta}]^2$ (ii) $f(\theta) = e^{\theta}\ln\theta,\ \theta > 0$.

* For students taking early transcendentals – although the transcendentals options are important, you should practice with **both** versions of this exam.

SOLUTIONS TO MIDTERM EXAMINATION #1

(15%) Q1. Find (a) $\lim_{x \to 2} \dfrac{\frac{1}{x^3} - \frac{1}{8}}{x - 2}$

(b) $\lim_{x \to 1^-} \dfrac{|x-1|}{x^2 - 5x + 4}$

(c) the domain of the function:

$$f(x) = \frac{\sqrt{6 - x^2}}{x - 2}$$

Solution

(a) Limit questions of this type can be approached in simple stages.

Stage 1

Take the value to which x tends, i.e. here x tends to 2, and evaluate the expression at that value; i.e. evaluate $\dfrac{\frac{1}{x^3} - \frac{1}{8}}{x - 2}$ at $x = 2$. We obtain $\dfrac{0}{0}$. This means that the limit, as posed, is 'in disguise' and we need to do some further investigation, i.e. $\dfrac{0}{0}$ gives us *no information* about the limit. (The other two possibilities are:

(i) $\dfrac{a}{0}$, $a \neq 0$, which means that the limit **doesn't exist**.

(ii) $\dfrac{b}{a}$, $a \neq 0$, which means that the limit **equals** $\dfrac{b}{a}$).

Stage 2

To further investigate the limit we 'play around' with the expression, e.g. simplify, factor, etc., using all the algebraic tools at our disposal (see Part 1).

$$\frac{\frac{1}{x^3} - \frac{1}{8}}{x - 2} = \frac{\frac{8 - x^3}{8x^3}}{x - 2} = \frac{8 - x^3}{8x^3(x - 2)} \qquad \text{(A)}$$

Stage 3

Repeat Stage 1 with (A) until we get something other than $\dfrac{0}{0}$. Here we *still* get $\dfrac{0}{0}$ when $x = 2$ in (A). So we continue to work on (A).
From (3.5.1), Part 1, $a^3 - b^3 = (a - b)(a^2 + ab + b^2)$.

With $a = 2,\ b = x:\ 8 - x^3 = (2-x)(4+2x+x^2)$. Hence

$$\frac{8-x^3}{8x^3(x-2)} = \frac{(2-x)(4+2x+x^2)}{8x^3(x-2)}$$
$$= \frac{-(x-2)(4+2x+x^2)}{8x^3(x-2)}$$
$$(\text{since } (2-x) = -(x-2))$$

Now, we can cancel the $(x-2)$ factors since when taking the limit, $x \neq 2$, i.e. x tends to 2 but never gets there!

$$\text{i.e.}\quad \lim_{x\to 2} \frac{8-x^3}{8x^3(x-2)} = \lim_{x\to 2}\left[-\frac{(4+2x+x^2)}{8x^3}\right] \qquad \text{(B)}$$

This time, letting $x = 2$ in the latest expression, i.e. (B), we obtain

$$-\frac{(4+4+4)}{8\cdot 8} = \frac{-12}{64} = \frac{-3}{16}$$

$$\text{i.e.}\quad \lim_{x\to 2}\frac{\frac{1}{x^3}-\frac{1}{8}}{x-2} = -\frac{3}{16}$$

We can summarize the process diagramatically as follows:

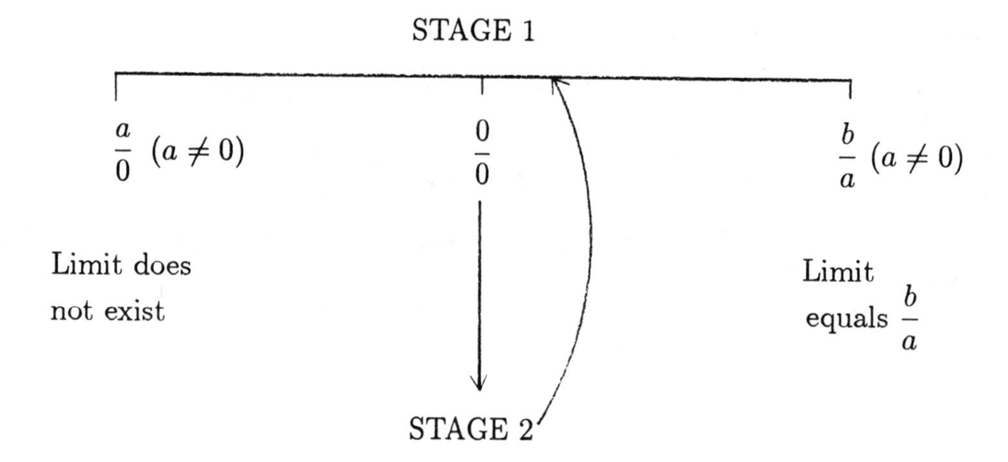

(b) **Stage 1**

$\left(\frac{0}{0}\right)$ (See (a))

Stage 2

$$|x-1| = \begin{cases} x-1, & x-1 \geq 0, \\ -(x-1), & x-1 < 0. \end{cases}$$

$$= \begin{cases} x-1, & x \geq 1, \\ 1-x, & x < 1. \end{cases} \qquad \text{(C)}$$

Also, $x^2 - 5x + 4 = (x-4)(x-1)$.

Hence

$$\lim_{x \to 1^-} \frac{|x-1|}{x^2 - 5x + 4} = \lim_{x \to 1^-} \frac{1-x}{(x-4)(x-1)}$$

(Since $x \to 1^-$, x approaches 1 from the left so that $x < 1$ always.

Hence $|x-1| = 1 - x$ from (C).)

Therefore

$$\lim_{x \to 1^-} \frac{|x-1|}{x^2 - 5x + 4} = \lim_{x \to 1^-} \left[-\frac{1}{(x-4)} \right] \qquad \text{(remember, } 1 - x = -(x-1)\text{)}$$

Stage 3

$x = 1$ in $-\dfrac{1}{x-4}$ gives $\dfrac{1}{3}$ which is the value of the limit.

(c) The domain is the set of values of x where $f(x)$ makes sense. Two things to remember (see Section 3.3(i)):

(i) The square root of a negative number 'does not make sense' in the real number system.

(ii) Division by zero does not make sense.

Clearly when $6 - x^2 < 0$ we have (i) and when $x = 2$ we have (ii). So our domain consists of values of x such that

$$6 - x^2 \geq 0, \quad x \neq 2$$

$$\text{i.e. } x^2 \leq 6, \quad x \neq 2$$

$$\text{i.e. } x \in (-\sqrt{6}, 2) \cup (2, \sqrt{6})$$

(20%) Q2. (a) Find $f'(x)$ if

(i) $f(x) = x^3(1 - x^3)^{\frac{4}{3}}$

(ii) $f(x) = \dfrac{(x^3 + 1)}{\csc x}$

(iii) $f(x) = \sqrt{\sin(\cos^2 x)}$

(b) Find the equation of the line tangent to the graph of

$$x^2y^2 = (y+1)^2(9-y^2)$$

at the point $(0, -3)$.

Solution

(a)(i) Product rule: $f'(x) = 3x^2(1-x^3)^{\frac{4}{3}} + x^3\dfrac{d}{dx}\left[(1-x^3)^{\frac{4}{3}}\right]$

To get $\dfrac{d}{dx}\left[(1-x^3)^{\frac{4}{3}}\right]$, write $u = (1-x^3)$.

Then

$$\begin{aligned}\frac{d}{dx}u^{\frac{4}{3}} \underset{\text{CHAIN RULE}}{=} \frac{d}{du}u^{\frac{4}{3}}\frac{du}{dx} &= \frac{4}{3}u^{\frac{1}{3}}\cdot(-3x^2)\\ &= -4x^2(1-x^3)^{\frac{1}{3}}\end{aligned}$$

Hence

$$f'(x) = 3x^2(1-x^3)^{\frac{4}{3}} - 4x^5(1-x^3)^{\frac{1}{3}}$$

(a)(ii) Quotient rule:

$$\begin{aligned}f'(x) &= \frac{3x^2\csc x - (-\cot x\csc x)(x^3+1)}{\csc^2 x}\\ &= \frac{3x^2 + (x^3+1)\cot x}{\csc^2 x}\end{aligned}$$

(a)(iii) Let $u = \sin(\cos^2 x)$.

$$\begin{aligned}\frac{d}{dx}\sqrt{\sin(\cos^2 x)} = \frac{d}{dx}\sqrt{u} = \frac{d}{du}\sqrt{u}\cdot\frac{du}{dx} &\quad \text{(chain rule)}\\ = \frac{1}{2\sqrt{u}}\cdot\frac{d}{dx}\sin(\cos^2 x)&\end{aligned}$$

To get $\dfrac{d}{dx}\sin(\cos^2 x)$, let $w = \cos^2 x$.

Then

$$\begin{aligned}\frac{d}{dx}\sin(\cos^2 x) = \frac{d}{dx}\sin w = \frac{d}{dw}\sin w\frac{dw}{dx} &\quad \text{(chain rule)}\\ = \cos w\cdot(-2\cos x\sin x)&\\ = -2\cos x\sin x\cos(\cos^2 x)&\end{aligned}$$

Hence,

$$\frac{d}{dx}\sqrt{\sin(\cos^2 x)} = \frac{1}{2\sqrt{\sin(\cos^2 x)}}$$

$$= \frac{-\sin 2x \cos(\cos^2 x)}{2\sqrt{\sin(\cos^2 x)}} \quad (\text{since } \sin 2x = 2\sin x \cos x)$$

(b) We need slope of the tangent at $(0, -3)$. Find y'.

Implicit differentiation w.r.t. x :

$$\underset{\text{(PRODUCT RULE)}}{2xy^2 + 2x^2yy'} = \underset{\text{(PRODUCT RULE + CHAIN RULE)}}{2(y+1)y'(9-y^2) - 2yy'(y+1)^2}$$

(RECALL $\frac{d}{dx}f(y) \underset{\text{CHAIN RULE}}{=} \frac{d}{dy}f(y)\frac{dy}{dx} = f'(y)\frac{dy}{dx} = f'(y)y'$,
e.g. $\frac{d}{dx}y^2 = 2yy'$.)
Hence,

$$y' = \frac{-2xy^2}{2x^2y - 2(y+1)(9-y^2) + 2y(y+1)^2}$$

$$= 0 \text{ at the point } (0, -3).$$

Thus, $y = -3$ is required tangent line.

(12%) Q3. Discuss the continuity of

$$f(x) = \begin{cases} x^2, & x < 0, \\ 2x - 1, & 0 \le x \le 2, \\ 2x^2 - 5, & x > 2. \end{cases}$$

Solution

$f(x)$ is continuous in the intervals $x < 0,\ 0 < x < 2,\ x > 2$ since it is represented by polynomials there.

The only suspect points are where $f(x)$ changes representation, i.e. $x = 0,\ 2$. Consider each in turn.

$\underline{x = 0}$

Continuity at $x = 0$ requires $\lim_{x\to 0} f(x) = f(0)$. Now, $f(0)$ is *defined* to be equal to -1 (i.e. let $x = 0$ in $2x - 1$).

For the existence of $\lim_{x\to 0} f(x)$, we must show $\lim_{x\to 0^-} f(x) = \lim_{x\to 0+} f(x)$.

Now, $\lim_{x\to 0^-} f(x) = \lim_{x\to 0^-} x^2$ (since $f(x) = x^2,\ x < 0$).

Therefore

$$\lim_{x\to 0^-} f(x) = 0.$$

Now,

$$\lim_{x\to 0+} f(x) = \lim_{x\to 0+} (2x - 1) \quad (\text{since} f(x) = 2x - 1, \text{ when } x \text{ is close to zero from the right})$$
$$= -1$$

Thus $\lim_{x\to 0^-} f(x) \neq \lim_{x\to 0+} f(x)$ so that $\lim_{x\to 0} f(x)$ *does not exist.*

So $f(x)$ is not continuous at $x = 0$.

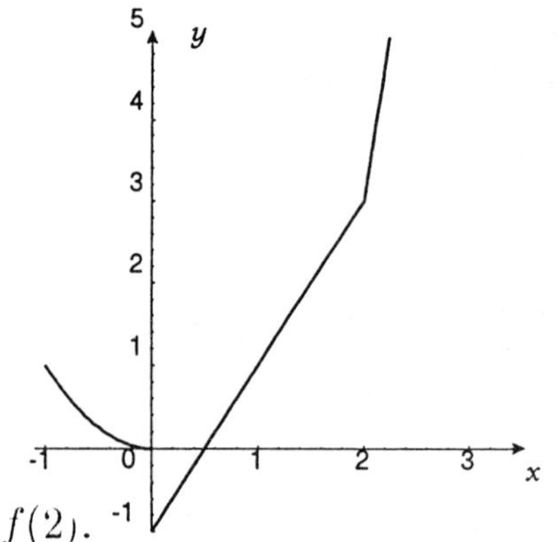

$\underline{x = 2}$

$f(2) = 3$ (let $x = 2$ in $2x - 1$)

$\lim_{x\to 2^-} f(x) = \lim_{x\to 2^-} (2x - 1) = 3$

$\lim_{x\to 2+} f(x) = \lim_{x\to 2+} (2x^2 - 5) = 3$

Hence $\lim_{x\to 2^-} f(x) = \lim_{x\to 2+} f(x) = \lim_{x\to 2} f(x) = 3 = f(2)$.

So $f(x)$ is continuous at $x = 2$.

(12%) Q4. Ship A is 10 miles north of ship B at 1300 hrs. Ship A sails east at 18 mph while ship B sails north at 12 mph. How fast is the distance between them changing at 1500 hrs?

Solution

Draw a diagram!

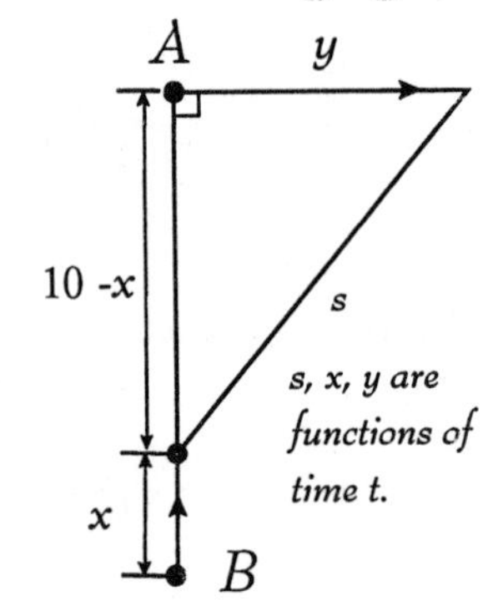

We know: $\dfrac{dy}{dt} = 18,\ \dfrac{dx}{dt} = 12$

We want: $\dfrac{ds}{dt}$

The key is to *relate* what we *want* to what we *know*. To do this, note that $s^2 = (10-x)^2 + y^2$. Differentiating both sides with respect to t (implicitly) we obtain

$$2s\frac{ds}{dt} = -2(10-x)\frac{dx}{dt} + 2y\frac{dy}{dt}. \qquad (1)$$

Now, after 2 hours, i.e. at 1500 hrs, $x = 24$ miles, $y = 36$ miles, so that $s^2 = 14^2 + 36^2 = 1492$, i.e. $s = \sqrt{1492}$.

Thus, from (1)

$$\left.\frac{ds}{dt}\right|_{t=1500\text{ hrs}} = \frac{1}{\sqrt{1492}}[(-14)(-12) + 36(18)]$$
$$= \frac{816}{\sqrt{1492}} \sim 21 \text{ mph}$$

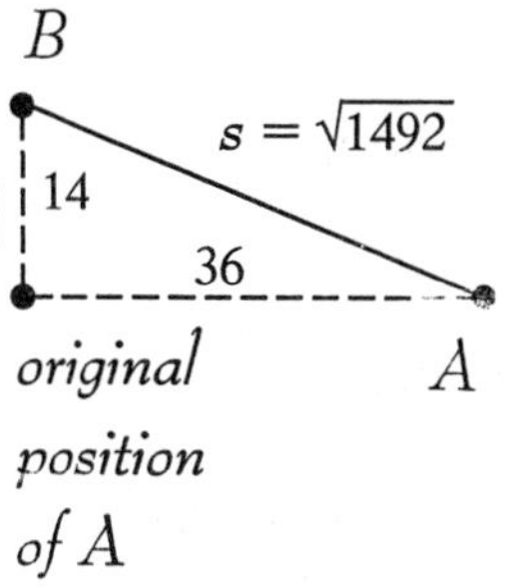

B

$s = \sqrt{1492}$

14

36

original position of A

A

(16%) Q5. (a) Use differentials to approximation $\sqrt{10}$.

(b) Show that Newton's method fails when applied to the equation $\sqrt[4]{x} = 0$ with any initial approximation $x_1 \neq 0$.

Solution

(a) Let $y = f(x) = \sqrt{x}$.

$$dy \simeq f(x_1 + \Delta x) - f(x_1) = \Delta y \quad \text{(standard formula)}$$

Hence, $f(x_1 + \Delta x) \simeq f(x_1) + dy$.

Let $x_1 = 9$, $\Delta x = dx = 1$. We want $f(10)$.

$$f(10) \simeq f(9) + dy$$

Now, $dy = \dfrac{1}{2\sqrt{x_1}}dx = \dfrac{1}{6}$.

Therefore

$$\sqrt{10} \simeq 3 + \frac{1}{6} = \frac{19}{6}$$

(b) For $f(x) = x^{\frac{1}{4}}$, $f'(x) = \dfrac{1}{4}x^{-\frac{3}{4}}$ and Newton's method gives:

$$x_{n+1} = x_n - \frac{f(x_n)}{f'(x_n)} = x_n - \frac{x_n^{\frac{1}{4}}}{\frac{1}{4}x_n^{-\frac{3}{4}}}$$
$$= x_n - 4x_n$$
$$= -3x_n.$$

Therefore each successive approximation becomes three times as large (in absolute value) as the previous one, so the sequence of approximations will not converge to the root (which is $x = 0$).

(25%) Q6. (a) Use the *limit definition* of the derivative to find $f'(x)$ if $f(x) = \sqrt{2x+3}$.

(b) What is the equation of the tangent line to the graph of $y = \sqrt{2x+3}$ at the point $(3, 3)$?

(c) What is $\frac{d^2}{dx^2}$ if $\sqrt{x} + \sqrt{y} = 321$?

Solution

(a) The limit definition gives:

$$\begin{aligned} f'(x) &= \lim_{h \to 0} \frac{f(x+h) - f(x)}{h} \\ &= \lim_{h \to 0} \frac{\sqrt{2(x+h)+3} - \sqrt{2x+3}}{h}. \end{aligned}$$

Following the procedure for limits in Q.1:

Stage 1

$\left(\frac{0}{0}\right)$

Stage 2

Rationalize numerator in expression (see Section 3.6, Part 1).

$$\begin{aligned} \frac{\sqrt{2(x+h)+3} - \sqrt{2x+3}}{h} \cdot \frac{\sqrt{2(x+h)+3} + \sqrt{2x+3}}{\sqrt{2(x+h)+3} + \sqrt{2x+3}} &= \frac{2(x+h)+3-2x-3}{h(\sqrt{2(x+h)+3} + \sqrt{2x+3})} \\ &= \frac{2h}{h(\sqrt{2(x+h)+3} + \sqrt{2x+3})} \qquad \text{(A)} \end{aligned}$$

Stage 3:

$$f'(x) = \lim_{h \to 0} \frac{2}{(\sqrt{2(x+h)+3} + \sqrt{2x+3})}$$

(Again, we can cancel the h's in (A) since $h \neq 0$ in the limit.)

Therefore

$$f'(x) = \frac{1}{\sqrt{2x+3}}$$

(b) We need the slope of the tangent line at $(3, 3)$, i.e. $y'(3)$.

From part (a), $y' = \dfrac{1}{\sqrt{2x+3}}$, therefore $y'(3) = \dfrac{1}{3}$.

Therefore tangent line:

$$y - 3 = \frac{1}{3}(x - 3)$$

$$(y - b = m(x - a) : \text{ line through } (a, b) \text{ with slope } m)$$

$$\text{i.e. } 3y = x + 6$$

(c) Implicit differentiation:

$$x^{\frac{1}{2}} + y^{\frac{1}{2}} = 321$$

$$\text{therefore } \quad \frac{d}{dx}\left[x^{\frac{1}{2}} + y^{\frac{1}{2}}\right] = 0$$

$$\text{therefore } \quad \frac{1}{2}x^{-\frac{1}{2}} + \frac{1}{2}y^{-\frac{1}{2}}y' = 0$$

$$\left(\frac{d}{dx}y^{\frac{1}{2}} = \frac{d}{dy}y^{\frac{1}{2}} \cdot \frac{dy}{dx} = \frac{1}{2}y^{-\frac{1}{2}} \cdot y'\right)$$

$$\text{therefore } \quad y' = -\frac{\sqrt{y}}{\sqrt{x}} \qquad \text{(B)}$$

Repeat implicit differentiation:

$$y'' = -\left[\frac{\frac{1}{2}y^{-\frac{1}{2}}y'\sqrt{x} - \frac{1}{2}x^{-\frac{1}{2}}\sqrt{y}}{x}\right].$$

But

$$y' = -\frac{\sqrt{y}}{\sqrt{x}} \quad \text{(from (B))}$$

$$\text{therefore } \quad y'' = -\left[\frac{\frac{1}{2\sqrt{y}} \cdot -\frac{\sqrt{y}}{\sqrt{x}}\sqrt{x} - \frac{1}{2}\frac{\sqrt{y}}{\sqrt{x}}}{x}\right]$$

$$= \frac{1}{2}\left[\frac{1 + \sqrt{\frac{y}{x}}}{x}\right]$$

$$= \frac{\sqrt{x} + \sqrt{y}}{2x^{\frac{3}{2}}}$$

SOLUTIONS TO MIDTERM EXAMINATION #2

(18%) Q1. Find each of the following limits:

(a) $\lim_{x \to 2} \frac{\sqrt{x+2}-2}{x-2}$

(b) $\lim_{x \to 1} \frac{|x-1|}{x^2-1}$

(c) $\lim_{x \to 0} \frac{\cos^2 4x - 1}{4\cos^2 5x - 4}$

Solution

(a) **Stage 1**

$$\left(\frac{0}{0}\right)$$

Stage 2

Rationalize the numerator (see Section 3.6, Part 1).

$$\frac{\sqrt{x+2}-2}{x-2} \cdot \frac{\sqrt{x+2}+2}{\sqrt{x+2}+2} = \frac{(x+2)-4}{(x-2)(\sqrt{x+2}+2)} = \frac{x-2}{(x-2)(\sqrt{x+2}+2)}$$

Hence

$$\begin{aligned}\lim_{x \to 2} \frac{\sqrt{x+2}-2}{x-2} &= \lim_{x \to 2} \frac{(x-2)}{(x-2)(\sqrt{x+2}+2)} \\ &= \lim_{x \to 2} \frac{1}{(\sqrt{x+2}+2)}\end{aligned}$$

Stage 3

Let $x = 2$ in expression:

$$\lim_{x \to 2} \frac{1}{\sqrt{x+2}+2} = \frac{1}{4}$$

(b) **Stage 1**

$$\left(\frac{0}{0}\right)$$

Stage 2

Simplify the expression.

$$|x-1| = \begin{cases} (x-1), & x-1 \geq 0, \quad \text{i.e. } x \geq 1 \\ -(x-1), & x-1 < 0, \quad \text{i.e. } x < 1 \end{cases}$$

Hence

$$\lim_{x \to 1^+} \frac{|x-1|}{x^2-1} = \lim_{x \to 1^+} \frac{(x-1)}{(x^2-1)} \quad (\text{since } x \geq 1 \text{ when } x \to 1^+)$$

$$= \lim_{x \to 1^+} \frac{1}{(x+1)} \qquad (A)$$

$$\lim_{x \to 1^-} \frac{|x-1|}{x^2-1} = \lim_{x \to 1^-} \frac{-(x-1)}{x^2-1} \quad (\text{since } \; x < 1 \text{ when } x \to 1^-)$$

$$= \lim_{x \to 1^-} -\frac{1}{(x+1)} \qquad (B)$$

Stage 3

$$\lim_{x \to 1^+} \frac{|x-1|}{x^2-1} = \lim_{x \to 1^+} \frac{1}{(x+1)} = \frac{1}{2} \quad (\text{using (A)}).$$

$$\lim_{x \to 1^-} \frac{|x-1|}{x^2-1} = \lim_{x \to 1^-} \frac{-1}{(x+1)} = -\frac{1}{2} \quad (\text{using (B)}).$$

Hence, both one-sided limits exist, **but are unequal**.

Thus $\lim_{x \to 1} \frac{|x-1|}{x^2-1}$ **does not exist.**

(c) **Stage 1**

$$\left(\frac{0}{0}\right)$$

Stage 2

Simplify the expression (see Section 6.2, Part 1).

$$\sin^2 \theta + \cos^2 \theta = 1 \qquad (A)$$

Hence

$$\cos^2 4x - 1 = -\sin^2 4x \quad (\text{letting } \theta = 4x \text{ in (A)})$$

and

$$4\cos^2 5x - 4 = 4(\cos^2 5x - 1)$$

$$= -4\sin^2 5x \quad (\theta = 5x \text{ in (A)})$$

Hence

$$\lim_{x\to 0} \frac{\cos^2 4x - 1}{4\cos^2 5x - 4} = \frac{1}{4} \lim_{x\to 0} \frac{\sin^2 4x}{\sin^2 5x} \qquad \text{(B)}$$

When dealing with trigonometric limits such as that in (B), try to use the standard results (see Appendix).

$$\lim_{\theta\to 0} \frac{\sin\theta}{\theta} = 1,$$

$$\lim_{\theta\to 0} \cos\theta = 1$$

$$\lim_{\theta\to 0} \sin\theta = 0.$$

i.e. try to make the '$\frac{\sin\theta}{\theta}$ shape' appear in the expression you're working with – this is a very common procedure in trigonometric limits.

i.e.

From (B)

$$\frac{1}{4} \lim_{x\to 0} \frac{\sin^2 4x}{\sin^2 5x} = \frac{1}{4} \lim_{x\to 0} \left(\frac{\sin 4x}{4x}\right)^2 \cdot (4x)^2 \cdot \left(\frac{5x}{\sin 5x}\right)^2 \cdot \frac{1}{(5x)^2}$$

(forcing the '$\frac{\sin\theta}{\theta}$ shape' to appear, but compensating for the 'extra terms' so that the expression remains the same)

$$= \frac{1}{4} \underbrace{\left(\lim_{x\to 0} \frac{\sin 4x}{4x}\right)^2}_{=1^2} \cdot \underbrace{\left(\lim_{x\to 0} \frac{5x}{\sin 5x}\right)^2}_{=1^2} \cdot \frac{4^2}{5^2}$$

Stage 3

$$\lim_{x\to 0} \frac{\cos^2 4x - 1}{4\cos^2 5x - 4} = \frac{1}{4} \cdot \frac{4^2}{5^2} = \frac{4}{25}$$

(20%) Q2. (a) Use the *limit definition* of the derivative to find $f'(x)$ if $f(x) = \frac{1}{3x+2}$.

(b) Find the equation of the tangent line to the curve $y = \frac{1}{3x+2}$ at the point $\left(-\frac{1}{3}, 1\right)$.

Solution

(a)

$$\begin{aligned}
f'(x) &= \lim_{h\to 0}\frac{f(x+h)-f(x)}{h}\\
&= \lim_{h\to 0}\frac{\frac{1}{3(x+h)+2}-\frac{1}{3x+2}}{h}\\
&= \lim_{h\to 0}\frac{3x+2-[3(x+h)+2]}{h[3(x+h)+2][3x+2]}\\
&= \lim_{h\to 0}\frac{-3h}{h(3(x+h)+2)(3x+2)}\\
&= \lim_{h\to 0}\frac{-3}{[3(x+h)+2][3x+2]} = \frac{-3}{(3x+2)^2}
\end{aligned}$$

(b) Slope at $\left(-\frac{1}{3}, 1\right)$ is

$$\begin{aligned}
f'\left(-\frac{1}{3}\right) &= \frac{-3}{\left(3\cdot\left(-\frac{1}{3}\right)+2\right)^2}\\
&= -3
\end{aligned}$$

Tangent line:

$$\begin{aligned}
y-1 &= -3\left(x+\frac{1}{3}\right)\\
\text{i.e.}\quad y &= -3x
\end{aligned}$$

(24%) Q3. In each of the following cases find $\frac{dy}{dx}$.

(a) $y = \dfrac{x^3+1}{x^2-1}$

(b) $y = (2x^3+1)(3x+1)^{\frac{1}{3}}$

(c) $y = \sqrt{1+\sin^2 5x}$

(d) $x^2 \tan y = x^3 + y^2$

Solution

(a) Quotient rule.

$$\frac{dy}{dx} = \frac{3x^2(x^2-1) - 2x(x^3+1)}{(x^2-1)^2} = \frac{x^4 - 3x^2 - 2x}{(x^2-1)^2}$$
$$= \frac{x(x^2 - 3x - 2)}{(x^2-1)^2}$$

(b) Product rule.

$$\frac{dy}{dx} = 6x^2(3x+1)^{\frac{1}{3}} + \frac{1}{3}(3x+1)^{-\frac{2}{3}} \cdot 3(2x^3+1)$$
$$= 6x^2(3x+1)^{\frac{1}{3}} + \frac{(2x^3+1)}{(3x+1)^{\frac{2}{3}}}$$

(c) Chain rule.

$$y = \sqrt{u}, \; u = 1 + \sin^2 5x$$
$$\frac{dy}{dx} = \frac{dy}{du}\frac{du}{dx} = \frac{1}{2}u^{-\frac{1}{2}} \cdot \frac{d}{dx}(\sin^2 5x)$$

To get $\frac{d}{dx}(\sin^2 5x)$, write $w = \sin 5x$.

$$\frac{d}{dx}w^2 = \frac{d}{dw}w^2\frac{dw}{dx} = 2w \cdot 5\cos 5x$$
$$= 10 \sin 5x \cos 5x$$
$$(= 5\sin 10x \text{ incidentally! i.e. } \sin 2\theta = 2\sin\theta\cos\theta)$$

Hence $\frac{dy}{dx} = \frac{5\sin 10x}{2\sqrt{1+\sin^2 5x}}$.

(d) Implicit differentiation.

$$\underbrace{2x\tan y + x^2\sec^2 y\frac{dy}{dx}}_{\text{(Product rule on } x^2\tan y)} = 3x^2 + 2y\frac{dy}{dx}$$

i.e.

$$\frac{dy}{dx}(x^2\sec^2 y - 2y) = 3x^2 - 2x\tan y$$
$$\text{therefore} \quad \frac{dy}{dx} = \frac{3x^2 - 2x\tan y}{x^2\sec^2 y - 2y}$$

(20%) Q4. (a) Define precisely what it means to say that a function $f(x)$ is continuous at $x = a$.

(b) Suppose

$$f(x) = \begin{cases} -x + 1, & x < -1, \\ x^2 + 1, & -1 < x \le 0, \\ x^2 + 2, & 0 < x \le 1, \\ 2, & x > 1, \end{cases}$$

(i) Sketch the graph of $y = f(x)$.

(ii) Indicate where $f(x)$ is not continuous – give reasons for your answers.

(iii) Is $f(x)$ continuous on $[0, 1]$? Explain.

Solution

(a) $f(x)$ is continuous at $x = a$ iff

$$\lim_{x \to a} f(x) = f(a) \qquad (*)$$

(b)(i)

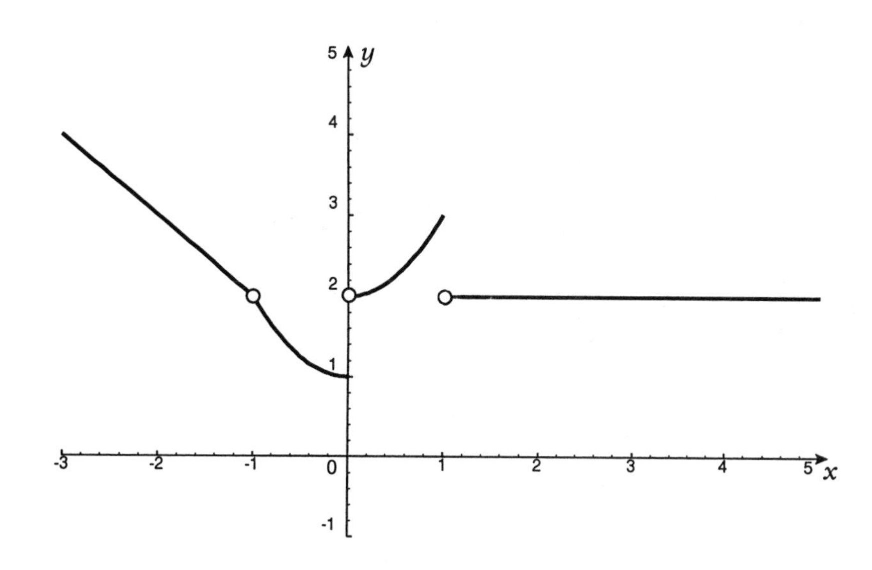

(b)(ii) It is clear *from the graph* that $f(x)$ is discontinuous at:

$x = -1$: since $f(-1)$ is *not defined* (hence (*) cannot possibly hold)

$x = 0$: since $\lim_{x \to 0^-} f(x) = 1 \neq \lim_{x \to 0+} f(x) = 2$. Hence $\lim_{x \to 0} f(x)$ does not exist so that (*) cannot possibly hold.

$x = 1$: since $\lim_{x \to 1^-} f(x) = 3 \neq \lim_{x \to 1+} f(x) = 2$. Hence $\lim_{x \to 1} f(x)$ does not exist so that (*) cannot possibly hold.

(b)(iii) $f(x)$ is not continuous on $[0, 1]$ since although it is continuous on $(0, 1)$ and continuous from the left at $x = 1$ i.e. $\lim_{x \to 1^-} f(x) = 3 = f(1)$, it is *not*

continuous from the right at $x = 0$ i.e. $\lim_{x \to 0+} f(x) = 2 \neq f(0) = 1$.

(18%) Q5. A water container has the shape of an inverted cone. Water is leaking out at a rate of $\frac{1}{2}$ m^3 per hour. The container has a height of 6 metres and the radius at the top of the container is 1 metre. How fast is the water level falling when the water in the container is 4 metres deep?

Solution

We know: $\frac{dV}{dt} = -\frac{1}{2}$ (V is volume)

$h =$ height $= 6m$

$r =$ radius at top $= 1m$

We want: $\frac{ds}{dt}$ when $s = 4$ (s is water level)

Diagram

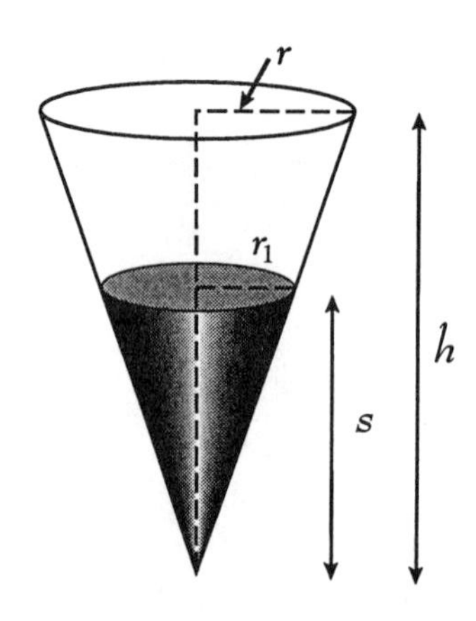

In order to relate what we *want* to what we *know,* we need a relation between s and V.

Now, volume of cone is $\frac{\pi}{3}r^2h$. Hence volume of water is $V = \frac{\pi}{3}r_1^2 s$ where r_1 is the radius at the top surface of the water. Now, by similar triangles, $\frac{r_1}{s} = \frac{r}{h} = \frac{1}{6}$.

Thus

$$\begin{aligned} V &= \frac{\pi}{3}\left(\frac{s}{6}\right)^2 s \\ &= \frac{\pi s^3}{108} \end{aligned}$$

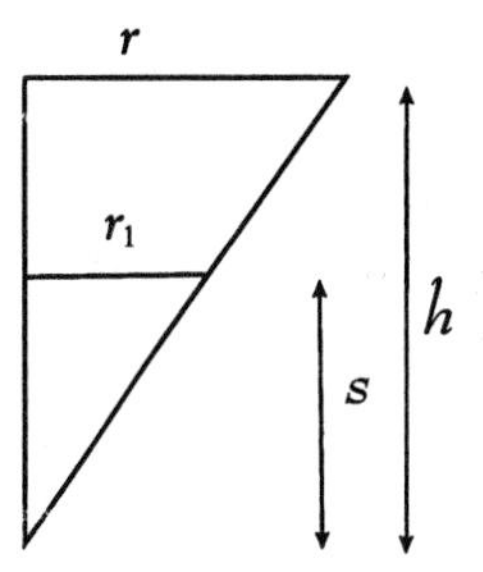

So

$$\frac{dV}{dt} = \frac{\pi}{108} \cdot 3s^2 \frac{ds}{dt} = -\frac{1}{2} \quad \text{(Given)}$$

Hence,

$$\begin{aligned} \frac{ds}{dt} &= \frac{-\frac{1}{2}}{\frac{\pi}{108} \cdot 3 \cdot (4)^2} \quad \text{when } s = 4 \\ &= -\frac{9}{8\pi} \text{ metres per hour} \end{aligned}$$

SOLUTIONS TO MIDTERM EXAMINATION #3

(17%) Q1. Find the following derivatives. You need not simplify your answer.

(a) If $y = \dfrac{\sin 2x}{\sqrt{x + \cos 3x}}$, find y'.

(b) If $y = \tan\left(\dfrac{1}{x^2}\right)$, find y', y''.

(c) If $x = \sin[\theta \tan\theta]^2$, find $\dfrac{dx}{d\theta}$.

Solution

(a) Quotient rule:

$$y' = \frac{2\cos 2x\sqrt{x+\cos 3x} - \sin 2x \frac{1}{2}(x+\cos 3x)^{-\frac{1}{2}}(1 - 3\sin 3x)}{(x+\cos 3x)}$$

(b) $y = \tan u$, $u = \dfrac{1}{x^2}$ (chain rule)

$$\begin{aligned}\frac{dy}{dx} = \frac{dy}{du}\frac{du}{dx} &= \sec^2 u \cdot \left(-\frac{2}{x^3}\right) \\ &= -\frac{2}{x^3}\sec^2\left(\frac{1}{x^2}\right)\end{aligned}$$

$$y'' = \frac{6}{x^4}\sec^2\left(\frac{1}{x^2}\right) - \frac{2}{x^3}\frac{d}{dx}\sec^2\left(\frac{1}{x^2}\right) \quad \text{(product rule)}$$

For $\dfrac{d}{dx}\sec^2\left(\dfrac{1}{x^2}\right)$, write $w = \dfrac{1}{x^2}$.

$$\begin{aligned}\frac{d}{dx}\sec^2 w = \frac{d}{dw}\sec^2 w\frac{dw}{dx} &= 2\sec w(\sec w\tan w)\left(-\frac{2}{x^3}\right) \\ &= \left(-\frac{4}{x^3}\right)\sec^2\left(\frac{1}{x^2}\right)\tan\left(\frac{1}{x^2}\right)\end{aligned}$$

Hence,

$$y'' = \frac{6}{x^4}\sec^2\left(\frac{1}{x^2}\right) - \frac{8}{x^6}\sec^2\left(\frac{1}{x^2}\right)\tan\left(\frac{1}{x^2}\right)$$

(c) $x = \sin u^2$ where $u = \theta \tan\theta$.

$$\frac{dx}{d\theta} = \frac{dx}{du}\frac{du}{d\theta} \qquad \text{(chain rule)}.$$

To get $\dfrac{dx}{du}$, apply chain rule once more:

$$x = \sin w \quad \text{where} \quad w = u^2.$$

$$\frac{dx}{du} = \frac{dx}{dw}\frac{dw}{du} = \cos w \cdot 2u = 2u\cos u^2.$$

Hence,

$$\begin{aligned}\frac{dx}{d\theta} &= 2u\cos u^2 \cdot \underbrace{\left[\tan\theta + \theta\sec^2\theta\right]}_{\text{PRODUCT RULE}} \\ &= 2\theta\tan\theta\cos(\theta\tan\theta)^2\left[\tan\theta + \theta\sec^2\theta\right].\end{aligned}$$

(17%) Q2. Evaluate the following limits. Show your work.

(a) $\displaystyle\lim_{x\to 0}\frac{1}{x}(\sqrt{2+x} - \sqrt{2-x})$

(b) $\displaystyle\lim_{x\to 0}(x^2+3x)\cot 6x$

(c) $\displaystyle\lim_{h\to 0}\frac{(1+h)^{-2} - 1^{-2}}{h}$

Solution

(a) **Stage 1**

$\left(\frac{0}{0}\right)$

Stage 2

Rationalize the numerator (see Section 3.6, Part 1):

$$\begin{aligned}&\frac{1}{x}(\sqrt{2+x} - \sqrt{2-x}) \cdot \left(\frac{\sqrt{2+x} + \sqrt{2-x}}{\sqrt{2+x} + \sqrt{2-x}}\right) \\ &= \frac{1}{x}\left[\frac{(2+x) - (2-x)}{\sqrt{2+x} + \sqrt{2-x}}\right] = \frac{2x}{x(\sqrt{2+x} + \sqrt{2-x})}\end{aligned}$$

Thus

$$\lim_{x\to 0}\frac{1}{x}(\sqrt{2+x} - \sqrt{2-x}) = \lim_{x\to 0}\frac{2}{\sqrt{2+x} + \sqrt{2-x}}$$

Stage 3

Let $x = 0$ in expression:

$$\lim_{x\to 0} \frac{2}{\sqrt{2+x}+\sqrt{2-x}} = \frac{1}{\sqrt{2}}$$

(b) **Stage 1**

$$\left(\frac{0}{0}\right)$$

Stage 2

Simplify:

$$\begin{aligned}\lim_{x\to 0}(x^2+3x)\cot 6x &= \lim_{x\to 0} x(x+3)\frac{\cos 6x}{\sin 6x} \\ &= \lim_{x\to 0}\frac{x}{\sin 6x}\cdot \lim_{x\to 0}(x+3)\cos 6x\end{aligned}$$

(Now begin looking for the appropriate 'shapes' – see Midterm #2, Q1 (c).)

$$\lim_{x\to 0}(x^2+3x)\cot 6x = \frac{1}{6}\underbrace{\lim_{x\to 0}\frac{6x}{\sin 6x}}_{=1}\cdot \underbrace{\lim_{x\to 0}(x+3)\cos 6x}_{=3\cdot 1}$$

Stage 3

$$\lim_{x\to 0}(x^2+3x)\cot 6x = \frac{1}{6}\cdot 1\cdot 3 = \frac{1}{2}$$

(c)

$$\lim_{h\to 0}\frac{(1+h)^{-2}-1^{-2}}{h} = \lim_{h\to 0}\frac{f(1+h)-f(1)}{h} = f'(1)$$

where $f(x) = x^{-2}$.

Therefore

$$f'(1) = \left[-\frac{2}{x^3}\right]_{x=1} = -2.$$

(Or go through stages 1 to 3, i.e.

Stage 1

$\left(\frac{0}{0}\right)$

Stage 2

Simplify:

$$\frac{(1+h)^{-2}-1^{-2}}{h} = \frac{\frac{1}{(1+h)^2}-1}{h}$$

Therefore

$$\begin{aligned}\lim_{h\to 0}\frac{(1+h)^{-2}-1^{-2}}{h} &= \lim_{h\to 0}\frac{1-(1+h)^2}{h(1+h)^2}\\ &= \lim_{h\to 0}\frac{-2h-h^2}{h(1+h)^2}\\ &= \lim_{h\to 0}\frac{-2-h}{(1+h)^2}\end{aligned}$$

Stage 3

$$\lim_{h\to 0}\frac{(1+h)^{-2}-1^{-2}}{h} = -2 \quad \text{as above}).$$

(11%) Q3. Let the function f be defined on $\left(-\frac{\pi}{2},\frac{\pi}{2}\right)$ by

$$f(x) = \begin{cases} \dfrac{\sin 6x}{\sin x}, & x \neq 0, \\ c, & x = 0, \end{cases}$$

where c is some constant. Find c such that f is a continuous function on $\left(-\frac{\pi}{2},\frac{\pi}{2}\right)$. Justify your conclusion.

Solution

The only suspect point as far as continuity is concerned is $x = 0$. At $x = 0$ we require $\lim_{x\to 0} f(x) = f(0) = c$ for continuity.

Now, $\lim_{x\to 0}\dfrac{\sin 6x}{\sin x}$ can be investigated as follows.

Stage 1

$\left(\frac{0}{0}\right)$

Stage 2

Simplify the expression with standard trigonometric limit results in mind.

i.e.

$$\begin{aligned}\lim_{x\to 0}\frac{\sin 6x}{\sin x} &= \lim_{x\to 0}\frac{\sin 6x}{6x}\cdot\frac{x}{\sin x}\cdot\frac{6x}{x}\\ &= \underbrace{\lim_{x\to 0}\frac{\sin 6x}{6x}}_{=1}\cdot\underbrace{\lim_{x\to 0}\frac{x}{\sin x}}_{=1}\cdot 6\end{aligned}$$

Stage 3

$$\lim_{x\to 0}\frac{\sin 6x}{\sin x} = 6$$

Hence $\lim_{x\to 0} f(x) = f(0) \Leftrightarrow 6 = c$

i.e. $c = 6$ will make $f(x)$ continuous on $\left(-\frac{\pi}{2}, \frac{\pi}{2}\right)$.

(11%) Q4. Find $f'(x)$ using the *limit definition* of a derivative if $f(x) = \dfrac{1}{\sqrt[3]{x}}$.

Solution

$$\begin{aligned}f'(x) &= \lim_{h\to 0}\frac{f(x+h)-f(x)}{h}\\ &= \lim_{h\to 0}\frac{\frac{1}{(x+h)^{\frac{1}{3}}}-\frac{1}{x^{\frac{1}{3}}}}{h}\end{aligned}$$

Stage 1

$$\left(\frac{0}{0}\right)$$

Stage 2

Simplify:

$$\frac{\frac{1}{(x+h)^{\frac{1}{3}}}-\frac{1}{x^{\frac{1}{3}}}}{h} = \frac{x^{\frac{1}{3}}-(x+h)^{\frac{1}{3}}}{hx^{\frac{1}{3}}(x+h)^{\frac{1}{3}}}$$

(Still gives $\left(\frac{0}{0}\right)$ so rationalize the numerator – see Example 3.3.7(viii), Part 1.)

Now, $a^3 - b^3 = (a-b)(a^2+ab+b^2)$ (See Section 3.5, Part 1.)

Let $a = x^{\frac{1}{3}}$, $b = (x+h)^{\frac{1}{3}}$.

Then

$$\frac{x^{\frac{1}{3}}-(x+h)^{\frac{1}{3}}}{hx^{\frac{1}{3}}(x+h)^{\frac{1}{3}}}\cdot\frac{x^{\frac{2}{3}}+x^{\frac{1}{3}}(x+h)^{\frac{1}{3}}+(x+h)^{\frac{2}{3}}}{x^{\frac{2}{3}}+x^{\frac{1}{3}}(x+h)^{\frac{1}{3}}+(x+h)^{\frac{2}{3}}}$$

simplifies to

$$\frac{\left(x^{\frac{1}{3}}\right)^3-\left((x+h)^{\frac{1}{3}}\right)^3}{hx^{\frac{1}{3}}(x+h)^{\frac{1}{3}}\left[x^{\frac{2}{3}}+x^{\frac{1}{3}}(x+h)^{\frac{1}{3}}+(x+h)^{\frac{2}{3}}\right]}$$

$$=\frac{-h}{hx^{\frac{1}{3}}(x+h)^{\frac{1}{3}}\left[x^{\frac{2}{3}}+x^{\frac{1}{3}}(x+h)^{\frac{1}{3}}+(x+h)^{\frac{2}{3}}\right]}$$

Stage 3

$$\lim_{h\to 0}\frac{\frac{1}{(x+h)^{\frac{1}{3}}}-\frac{1}{x^{\frac{1}{3}}}}{h}=\lim_{h\to 0}\frac{-1}{x^{\frac{1}{3}}(x+h)^{\frac{1}{3}}\left(x^{\frac{2}{3}}+x^{\frac{1}{3}}(x+h)^{\frac{1}{3}}+(x+h)^{\frac{2}{3}}\right)}$$

$$=-\frac{1}{x^{\frac{1}{3}}\cdot x^{\frac{1}{3}}\cdot\left(x^{\frac{2}{3}}+x^{\frac{2}{3}}+x^{\frac{2}{3}}\right)}$$

$$=-\frac{1}{3x^{\frac{4}{3}}}$$

(15%) Q5. (a) Determine all points where the graph of the equation

$$\sin x\cos y+(1-x^2)y^2=\pi^2$$

intersects the y-axis.

(b) Find the equation of the tangent line to the graph at each of the points found in (a).

Solution

(a) Intersection on y-axis when $x=0$, i.e. $y^2=\pi^2$, i.e. $y=\pm\pi$.

(b) First get y' for slope.

Implicit differentiation:

$$\cos x\cos y-\sin x(\sin y\cdot y')+(-2xy^2+2yy'(1-x^2)=0.$$

Let $x = 0$ (to get slope of tangent at the points where graph intersects y-axis)

$$\cos y + 2yy' = 0 \quad \text{i.e. } y' = -\frac{\cos y}{2y}$$

Now let $y = \pm\pi$ so that

$$\begin{aligned} y' &= -\frac{\cos(\pm\pi)}{2(\pm\pi)} \\ &= -\frac{(-1)}{\pm 2\pi} = \pm\frac{1}{2\pi} \end{aligned}$$

Tangent line at $(0, \pi)$ is:

$$y - \pi = \frac{1}{2\pi}(x) \quad \text{i.e. } y = \frac{x}{2\pi} + \pi$$

Tangent line at $(0, -\pi)$ is:

$$y + \pi = -\frac{1}{2\pi}(x) \quad \text{i.e. } y = -\left(\frac{x}{2\pi} + \pi\right)$$

(11%) Q6. (a) At a certain instant, the surface area S of a spherical balloon is increasing at the same rate as its radius r is increasing. What is the radius at that instant?

(b) A spherical raindrop accumulates moisture (through condensation) at a rate proportional to its surface area S. Prove that the radius r increases at a constant rate.

Solution

(a) At the instant $t = t_0$, we know that $\frac{dS}{dt} = \frac{dr}{dt} > 0$, where $S = 4\pi r^2$.
Hence at $t = t_0$,

$$\begin{aligned} 8\pi r\frac{dr}{dt} &= \frac{dr}{dt} \\ \frac{dr}{dt}(8\pi r - 1) &= 0 \\ r &= \frac{1}{8\pi} \end{aligned}$$

(b) We know that $\frac{dV}{dt} = kS$ where the volume $V = \frac{4}{3}\pi r^3$, $S = 4\pi r^2$ and $k = \text{constant} > 0$. Hence, $4\pi r^2\frac{dr}{dt} = 4\pi r^2 k$. Consequently $\frac{dr}{dt} = k = \text{constant} > 0$.

(16%) Q7. (a) The kinetic energy K of a mass m moving with speed v is given by

$$K = \frac{1}{2}mv^2.$$

Use differentials to estimate the percentage increase in kinetic energy of a mass m if its speed is decreased from 160 km/h to 157 km/h.

(b) Use differentials to estimate $\sin(58°)$.

OR*

Sketch the curve with polar equation $r = 2\cos\theta + 2\sin\theta$.

Solution

(a)

$$dK = 2 \cdot \frac{1}{2}mvdv = mvdv$$

If $v = 160,\ dv = -3,$

$$\begin{aligned} dK = 160\text{m}(-3) &= -480\text{m} \\ &= \text{change in K.E.} \end{aligned}$$

As a percentage,

$$\begin{aligned} \text{change in kinetic energy} &= \frac{-480\text{m}}{\frac{1}{2}\text{m}(160)^2} \cdot 100 \\ &= -3.75\% \end{aligned}$$

(b)

$$y = f(x) = \sin x \Rightarrow dy = \cos x dx$$

When $x = \frac{\pi}{3},\ dx = -\frac{2\pi}{180} = -\frac{\pi}{90}$ (i.e. $-2°$)

$$dy = \cos\left(\frac{\pi}{3}\right)\left(-\frac{\pi}{90}\right) = -\frac{1}{2}\frac{\pi}{90} = -\frac{\pi}{180}$$

So

$$\begin{aligned} \sin(58°) = f\left(\frac{58\pi}{180}\right) &\sim f\left(\frac{\pi}{3}\right) + dy \\ &= \frac{\sqrt{3}}{2} - \frac{\pi}{180} \\ &\sim 0.8486 \end{aligned}$$

OR*

$$
\begin{aligned}
&r = 2\cos\theta + 2\sin\theta\\
&\quad\Leftrightarrow r^2 = 2r\cos\theta + 2r\sin\theta\\
&\quad\Leftrightarrow x^2 + y^2 = 2x + 2y \qquad \text{(Polar to Cartesian – see (4.1.2), Part 1)}\\
&\quad\Leftrightarrow (x^2 - 2x) + (y^2 - 2y) = 0\\
&\quad\Leftrightarrow (x-1)^2 + (y-1)^2 = 2 \qquad \text{(Completing the square – see Note 3.7.5, Part 1)}
\end{aligned}
$$

This is the standard equation for a circle center $(1, 1)$ and radius $\sqrt{2}$ in the cartesian coordinate system (see (4.1.5), Part 1).

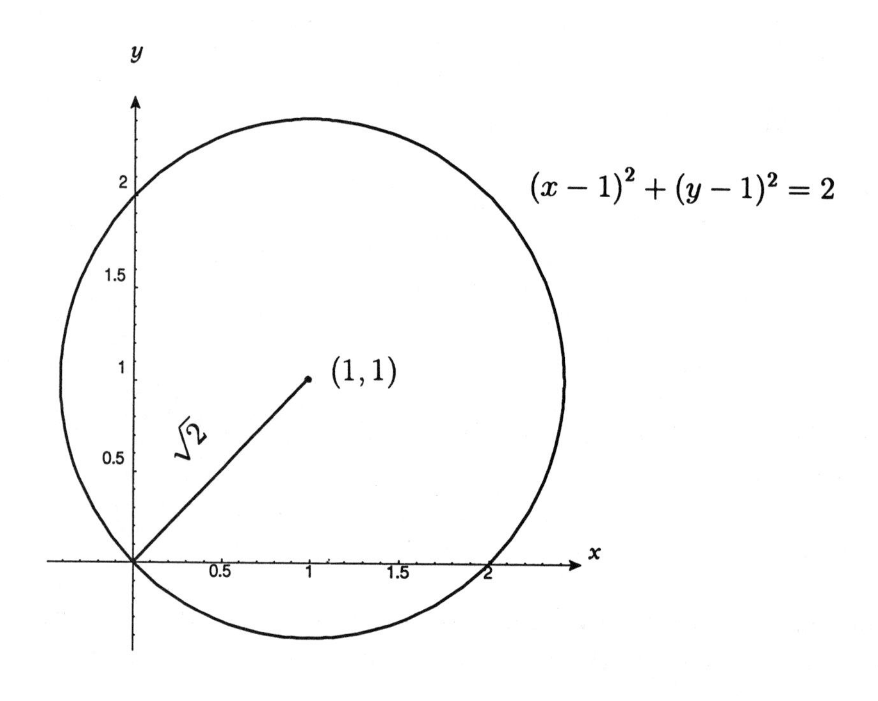

SOLUTIONS TO MIDTERM EXAMINATION #4

(14%) Q1. (a) Let $f(x) = \dfrac{x}{x+2}$ and $g(x) = \dfrac{\sqrt{x-1}}{x}$. Find $(f \circ g)(x)$ and its domain.

(b) If $f(x) = \cos\left[1 + \sqrt{x}\sin(x^3+4)\right]$, find $f'(x)$. (You need not simplify your answer.)

Solution

(a) Domain of f is $x \in \mathbb{R}\backslash\{-2\}$ (since $f(x)$ is not defined at $x = -2$)

Domain of g is $x \in \mathbb{R}:$ $\underbrace{x - 1 \geq 0}_{\text{ensures we take square root of a positive number}}$, $\underbrace{x \neq 0}_{\text{avoids division by zero}}$

i.e. $x \in \mathbb{R} : x \geq 1$

$$(f \circ g)(x) = f(g(x)) = \frac{g(x)}{g(x)+2}, \quad x \geq 1 \quad \text{(since domain of } g \text{ is } x \geq 1)$$
$$= \frac{\frac{\sqrt{x-1}}{x}}{\frac{\sqrt{x-1}}{x} + 2}, \quad x \geq 1$$

Domain of $f \circ g(x)$ requires first that $g(x)$ be defined, i.e. $x \geq 1$ and second that $f \circ g(x)$ be defined.

Now $(f \circ g)(x) = \dfrac{\sqrt{x-1}}{\sqrt{x-1}+2x}$, $x \geq 1$.

The denominator is never zero (since $4x^2 - x + 1 = 0$ has no real roots) so that the domain of $(f \circ g)(x)$ remains $x \in \mathbb{R} : x \geq 1$.

(b) Let $f(x) = y = \cos u$ where $u = 1 + \sqrt{x}\sin(x^3+4)$

$$f'(x) = \frac{dy}{dx} = \frac{dy}{du}\frac{du}{dx} = -\sin u \frac{d}{dx}\left[1 + \sqrt{x}\sin(x^3+4)\right]$$
$$= -\sin u\left[\frac{1}{2}x^{-\frac{1}{2}}\sin\left(x^3+4\right) + \sqrt{x}\cdot 3x^2\cos(x^3+4)\right]$$
$$= -\sin\left(1+\sqrt{x}\sin(x^3+4)\right)\left[\frac{1}{2\sqrt{x}}\sin(x^3+4) + 3x^{\frac{5}{2}}\cos(x^3+4)\right]$$

(14%) Q2. (a) Show that the sum of the x- and y-intercepts of any tangent line to the curve $\sqrt{x} - \sqrt{y} = a$, is equal to a^2.

(b) Find $\dfrac{d^2y}{dx^2}$ if $y = \sqrt[3]{1 + \cos x}$. (You need not simplify your answer.)

Solution

(a) To find tangent line we need slope at some point (x_1, y_1) on the curve. Implicit differentiation:

$$\frac{1}{2\sqrt{x}} - \frac{1}{2\sqrt{y}}y' = 0 \quad \text{i.e. } y' = \frac{\sqrt{y}}{\sqrt{x}}$$

Tangent line at (x_1, y_1)

$$y - y_1 = \frac{\sqrt{y_1}}{\sqrt{x_1}}(x - x_1)$$

x-intercept$(y = 0)$: $x = -y_1\dfrac{\sqrt{x_1}}{\sqrt{y_1}} + x_1 = x_1 - \sqrt{x_1}\sqrt{y_1}$

y-intercept$(x = 0)$: $y = -x_1\frac{\sqrt{y_1}}{\sqrt{x_1}} + y_1 = y_1 - \sqrt{x_1}\sqrt{y_1}$

Sum the intercepts:

$$\begin{aligned}\text{Sum } &= (x_1 - \sqrt{x_1}\sqrt{y_1}) + (y_1 - \sqrt{x_1}\sqrt{y_1}) \\ &= x_1 - 2\sqrt{x_1}\sqrt{y_1} + y_1 = (\sqrt{x_1} - \sqrt{y_1})^2 \\ &= (a)^2 = a^2\end{aligned}$$

(b) $y = u^{\frac{1}{3}}$, $u = 1 + \cos x$

$$\begin{aligned}\frac{dy}{dx} &= \frac{dy}{du}\frac{du}{dx} = \frac{1}{3}u^{-\frac{2}{3}}(-\sin x) \\ &= -\frac{\sin x}{3(1 + \cos x)^{\frac{2}{3}}}\end{aligned}$$

$$\frac{d^2y}{dx^2} = \frac{-\cos x \cdot 3(1 + \cos x)^{\frac{2}{3}} + 3\sin x \cdot \frac{2}{3}(1 + \cos x)^{-\frac{1}{3}}(-\sin x)}{3(1 + \cos x)^{\frac{4}{3}}}$$

(14%) Q3. Find $f'(x)$ using the *limit definition* of the derivative if

$$f(x) = \frac{1}{\sqrt{x-a}} \qquad (a \text{ is a constant}).$$

State the domain of the function and of its derivative.

Solution

$$\begin{aligned} f'(x) &= \lim_{h\to 0} \frac{f(x+h) - f(x)}{h} \\ &= \lim_{h\to 0} \frac{\frac{1}{\sqrt{x+h-a}} - \frac{1}{\sqrt{x-a}}}{h} \end{aligned}$$

Stage 1

$\left(\frac{0}{0}\right)$

Stage 2

Simplify and rationalize the numerator.

$$\begin{aligned} \frac{\frac{1}{\sqrt{x+h-a}} - \frac{1}{\sqrt{x-a}}}{h} &= \frac{\sqrt{x-a} - \sqrt{x+h-a}}{h\sqrt{x+h-a}\sqrt{x-a}} \\ &= \frac{\sqrt{x-a} - \sqrt{x+h-a}}{h\sqrt{x+h-a}\sqrt{x-a}} \cdot \frac{\sqrt{x-a} + \sqrt{x+h-a}}{\sqrt{x-a} + \sqrt{x+h-a}} \\ &= \frac{(x-a) - (x+h-a)}{h\sqrt{x+h-a}\sqrt{x-a}(\sqrt{x-a} + \sqrt{x+h-a})} \end{aligned}$$

Hence

$$f'(x) = \lim_{h\to 0} \frac{-1}{\sqrt{x+h-a}\sqrt{x-a}(\sqrt{x-a} + \sqrt{x+h-a})}$$

Stage 3

$$f'(x) = -\frac{1}{2(x-a)^{\frac{3}{2}}}$$

$$\left.\begin{array}{l} \text{Domain of } f \text{ is } x \in \mathbb{R}: \ x > a \\ \text{Domain of } f' \text{ is } x \in \mathbb{R}: \ x > a \end{array}\right\} \text{ ensures: } \begin{array}{l} \text{(i) No division by zero} \\ \text{(ii) Square roots are applied to} \\ \quad \text{positive numbers only} \end{array}$$

(16%) Q4. Evaluate each limit or explain why there is no limit.

(a) $\lim_{x \to 0} \frac{2 - \sqrt{1 - x^2}}{x}$

(b) $\lim_{x \to 0} \frac{\tan^2(8x)}{3x}$

(c) $\lim_{x \to 1^-} \left[\frac{(x^2 - 2x + 1)^{-1} - (x - 1)^{-1}}{(x^2 - 9)} \right]$

Solution

(a) **Stage 1**

$$\left(\frac{1}{0}\right)$$

Need go no further. Limit doesn't exist, i.e. as $x \to 0$ the denominator $\to 0$ while the numerator $\to 1$ so the expression grows larger and larger.

(b) **Stage 1**

$$\left(\frac{0}{0}\right)$$

Stage 2

Simplify

$$\begin{aligned}
\frac{\tan^2(8x)}{3x} &= \frac{1}{3} \frac{\sin^2 8x}{x \cos^2 8x} \\
&= \frac{1}{3} \left(\frac{\sin 8x}{8x} \right)^2 \cdot (8x)^2 \cdot \left(\frac{1}{\cos 8x} \right)^2 \cdot \frac{1}{x} \\
&= \frac{64x}{3} \cdot \left(\frac{\sin 8x}{8x} \right)^2 \left(\frac{1}{\cos 8x} \right)^2
\end{aligned}$$

Stage 3

$$\begin{aligned}
\lim_{x \to 0} \frac{\tan^2(8x)}{3x} &= \underbrace{\lim_{x \to 0} \left(\frac{64}{3} x \right)}_{=0} \cdot \underbrace{\lim_{x \to 0} \left(\frac{\sin 8x}{8x} \right)^2}_{=1} \cdot \underbrace{\lim_{x \to 0} \left(\frac{1}{\cos 8x} \right)^2}_{=1} \\
&= 0
\end{aligned}$$

(c) **Stage 1**

$\left(\dfrac{\frac{1}{0}-\frac{1}{0}}{-8}\right)$ Doesn't mean anything so go to Stage 2.

Stage 2

Simplify expression:

$$\lim_{x\to 1^-}\left[\frac{(x^2-2x+1)^{-1}-(x-1)^{-1}}{(x^2-9)}\right] = \lim_{x\to 1^-}\left[\frac{\left[(x-1)^2\right]^{-1}-(x-1)^{-1}}{(x^2-9)}\right]$$

i.e.

$$\begin{aligned}\lim_{x\to 1^-}\left[\frac{(x-1)^{-2}-(x-1)^{-1}}{(x^2-9)}\right] &= \lim_{x\to 1^-}\left[\frac{(x-1)^{-2}\left[1-(x-1)\right]}{(x^2-9)}\right]\\ &= \lim_{x\to 1^-}\frac{(2-x)}{(x-1)^2(x^2-9)}\end{aligned}$$

Stage 3

$\left(\dfrac{1}{0}\right)$ so limit doesn't exist since the numerator tends to 1 while the denominator gets smaller and smaller – hence the expression grows indefinitely.

(14%) Q5. Consider the function

$$f(x)=\begin{cases} x^2-x+1, & x<1,\\ 1, & x\geq 1.\end{cases}$$

(a) Is $f(x)$ continuous at $x=1$? Why?

(b) Is $f(x)$ differentiable at $x=1$? Why?

Solution

(a) $f(x)$ is continuous at $x=1$ iff

$$\lim_{x\to 1} f(x) = f(1).$$

Now,

$$\lim_{x\to 1^-} f(x) = \lim_{x\to 1^-} (x^2 - x + 1) = 1$$
$$\lim_{x\to 1^+} f(x) = \lim_{x\to 1^+} 1 = 1$$

$$\text{Therefore} \quad \lim_{x\to 1} f(x) = 1 = f(1)$$

Hence f is continuous at $x = 1$.

(b) $f(x)$ is differentiable at $x = 1$ iff the limit

$$\lim_{h\to 0} \frac{f(1+h) - f(1)}{h} \quad \text{exists.}$$

Now,

$$\begin{aligned}
\lim_{h\to 0^-} \frac{f(1+h) - f(1)}{h} &= \lim_{h\to 0^-} \frac{(1+h)^2 - (1+h) + 1 - 1}{h} \\
&= \lim_{h\to 0^-} \frac{h^2 + h}{h} \\
&= \lim_{h\to 0^-} 1 + h = 1
\end{aligned}$$

and

$$\begin{aligned}
\lim_{h\to 0+} \frac{f(1+h) - f(1)}{h} &= \lim_{h\to 0+} \frac{1-1}{h} \\
&= \lim_{h\to 0+} \frac{0}{h} \\
&= \lim_{h\to 0+} 0 = 0.
\end{aligned}$$

Hence

$$\lim_{h\to 0+} \frac{f(1+h) - f(1)}{h} \neq \lim_{h\to 0^-} \frac{f(1+h) - f(1)}{h}$$

so $\lim_{h\to 0} \frac{f(1+h) - f(1)}{h}$ doesn't exist.

i.e. f is *not* differentiable at $x = 1$.

(14%) Q6. A right circular cone of radius 2m and height 4m has its tip oriented downward and is filled with oil. The oil is escaping at a rate of 5cm^3/sec. How fast is the area of the top surface of the oil changing when the height of oil in the cone is 600cm?

Solution

Diagram:

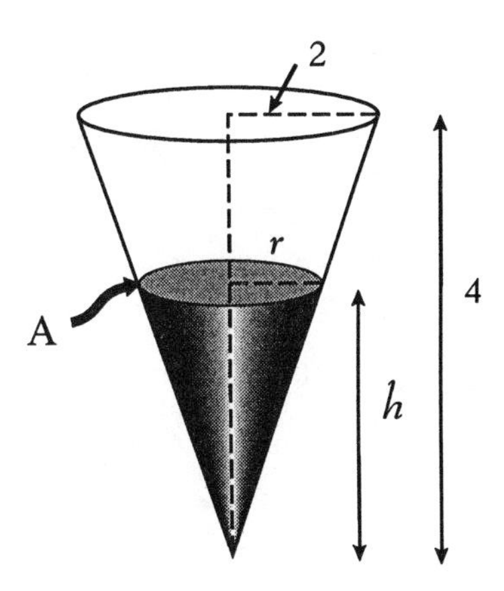

We know:

$$\frac{dV}{dt} = -5, \ V \text{ is volume of oil}$$

We want:

$$\frac{dA}{dt} \text{ when } h = 600 \ (A \text{ is area of top surface of oil})$$

Now, $V = \frac{\pi}{3} r^2 h, \ A = \pi r^2$.

By similar triangles: $\frac{r}{h} = \frac{2}{4}$ i.e. $h = 2r$.

Hence $V = \frac{2\pi}{3} r^3 = \frac{2\pi}{3} \left(\frac{A}{\pi}\right)^{\frac{3}{2}} = \frac{2}{3} \frac{A^{\frac{3}{2}}}{\sqrt{\pi}}$

Relate what we *want* to what we *know*:

$$-5 = \frac{dV}{dt} = \frac{2}{3\pi} \cdot \frac{3}{2} A^{\frac{1}{2}} \cdot \frac{dA}{dt} = \frac{1}{\sqrt{\pi}} A^{\frac{1}{2}} \frac{dA}{dt}.$$

Now, when $h = 600, \ r = \frac{h}{2} = 300, \ A = \pi(300)^2$.

Hence $-5 = \frac{1}{\sqrt{\pi}} \sqrt{\pi} \cdot 300 \frac{dA}{dt}$

$$\frac{dA}{dt} = -\frac{1}{60} \text{cm}^2/\text{sec}.$$

(14%) Q7. (a) If $f(x) = |x^3 - x^2|$, find f' and f''. What are their domains?

(b) Does the equation

$$x^8 + 3x^2 - x - 2 = 0$$

have any real roots?

OR*

Sketch the curve with equation $(x^2 + y^2)^6 = 16x^4y^4$.

Solution

(a)

$$f(x) = \begin{cases} x^3 - x^2, & x^3 - x^2 \geq 0, \text{ i.e. } x \geq 1 \\ x^2 - x^3, & x^3 - x^2 < 0, \text{ i.e. } x < 1 \end{cases}$$ (See Section 3.8, Part 1)

$$f'(x) = \begin{cases} 3x^2 - 2x, & x > 1 \\ 2x - 3x^2, & x < 1. \end{cases}$$

$$f''(x) = \begin{cases} 6x - 2, & x > 1, \\ 2 - 6x, & x < 1. \end{cases}$$

Domain of f is $\{x : x \in \mathbb{R}\}$
Domain of f' is $\{x : x \in \mathbb{R}\backslash\{1\}\}$
Domain of f'' is $\{x : x \in \mathbb{R}\backslash\{1\}\}$

(The function is not differentiable at $x = 1$ i.e.

$$\left.\begin{array}{l} \lim\limits_{h\to 0+} \dfrac{f(1+h) - f(1)}{h} = 1 \\ \lim\limits_{h\to 0-} \dfrac{f(1+h) - f(1)}{h} = -1 \end{array}\right\}$$

Not equal so $f'(1) = \lim\limits_{h\to 0} \dfrac{f(1+h) - f(1)}{h}$ doesn't exist!)

(b) Let $f(x) = x^8 + 3x^2 - x - 2$.

Let $x = 0$, $f(0) = -2$.

Let $x = 1$, $f(1) = 1$.

Since f is continuous on $[0, 1]$ (since f is a polynomial) by the Intermediate Value Theorem, $\exists\ c \in (0, 1)$ such that $f(c) = 0$, i.e. $f(x) = 0$ has at least one real root $x = c$ in the interval $(0, 1)$.

(In other words, the graph of the function crosses the x-axis somewhere between $x = 0$ and $x = 1$. This intersection is a root of the equation $f(x) = 0$).

OR*

Let $r^2 = x^2 + y^2,\ x = r\cos\theta,\ y = r\sin\theta$.
Cartesian form of equation now assumes the following polar form:

$$
\begin{aligned}
(r^2)^6 &= 16(r\cos\theta)^4(r\sin\theta)^4 \\
\text{i.e.}\quad r^{12} &= 16r^4\cos^4\theta r^4\sin^4\theta \\
\text{i.e.}\quad r^4 &= 16\cos^4\theta\sin^4\theta \\
&= (2\cos\theta\sin\theta)^4 \\
&= (\sin 2\theta)^4. \qquad \text{(See Section 6.2, Part 1)}
\end{aligned}
$$

Hence,

$$
r = |\sin 2\theta| = \begin{cases} \sin 2\theta, & \sin 2\theta \geq 0, \\ -\sin 2\theta, & \sin 2\theta < 0. \end{cases}
$$

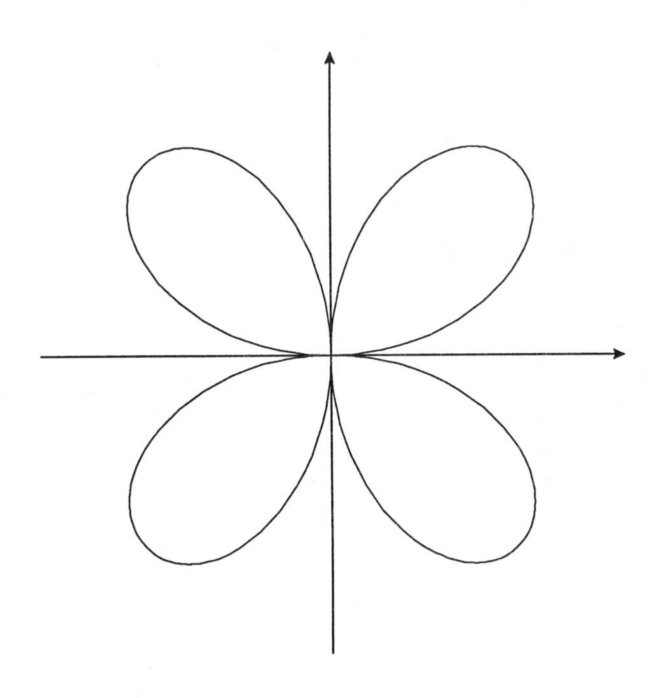

SOLUTIONS TO MIDTERM EXAMINATION #5

(25%) Q1. Find the limit if it exists. Explain why there is no limit otherwise.

(a) $\lim\limits_{x \to 4} \dfrac{\sqrt{x}}{2 + 2x - x^2}$

(b) $\lim\limits_{x \to -5} \dfrac{1}{x + 5}$

(c) $\lim\limits_{x \to 0} \dfrac{|x|}{2x}$

(d) $\lim\limits_{x \to 0} \dfrac{x^2 + \sin x}{\sqrt{x + 4} - 2}$

Solution

(a) **Stage 1**

$\left(\dfrac{2}{-6}\right)$

i.e.

$$\lim_{x \to 4} \frac{\sqrt{x}}{2 + 2x - x^2} = -\frac{1}{3}$$

$\left(f(x) = \dfrac{\sqrt{x}}{2 + 2x - x^2} \text{ is continuous at } x = 4 \right)$

(b) **Stage 1**

$\left(\dfrac{1}{0}\right)$

Limit doesn't exist. Denominator gets smaller while numerator stays constant. Hence expression grows indefinitely.

(c) **Stage 1**

$\left(\dfrac{0}{0}\right)$

Stage 2

Simplify the expression:

$$\frac{|x|}{2x} = \begin{cases} \dfrac{x}{2x}, & x > 0 \\ -\dfrac{x}{2x}, & x < 0 \end{cases} = \begin{cases} \dfrac{1}{2}, & x > 0 \\ -\dfrac{1}{2}, & x < 0 \end{cases}.$$

(Note that the expression is not defined at $x = 0$ – it doesn't have to be for the limit to exist!)

Stage 3

$$\lim_{x \to 0-} \frac{|x|}{2x} = -\frac{1}{2}; \quad \lim_{x \to 0+} \frac{|x|}{2x} = \frac{1}{2}$$

Hence both one-sided limits exist but the fact that they are not equal means that $\lim_{x \to 0} \frac{|x|}{2x}$ *does not exist.*

(d) **Stage 1**

$\left(\frac{0}{0}\right)$

Stage 2

Simplify expression:

$$\frac{x^2 + \sin x}{\sqrt{x+4} - 2} \cdot \frac{\sqrt{x+4} + 2}{\sqrt{x+4} + 2} = \frac{(x^2 + \sin x)(\sqrt{x+4} + 2)}{x}$$

Stage 3

$$\begin{aligned} &\lim_{x \to 0} \frac{(x^2 + \sin x)(\sqrt{x+4} + 2)}{x} \\ &= \lim_{x \to 0} \left(x + \frac{\sin x}{x}\right)(\sqrt{x+4} + 2) \\ &= \left[\lim_{x \to 0} x + \lim_{x \to 0} \frac{\sin x}{x}\right] \lim_{x \to 0} (\sqrt{x+4} + 2) \\ &= (0 + 1)(4) = 4 \end{aligned}$$

(25%) Q2. (a) Is the following function continuous at $x = 3$? Give reasons for your answer.

$$f(x) = \begin{cases} \dfrac{x^2 - x - 6}{x - 3}, & x \neq 3, \\ 3, & x = 3. \end{cases}$$

Sketch the graph of the function.

(b) Is the following function continuous at $x = -2$? Give reasons for your answer.

$$f(x) = \frac{x^2 - 4}{x + 2}.$$

Solution

(a)

$$f(x) = \begin{cases} \dfrac{(x-3)(x+2)}{(x-3)}, \; x \neq 3, \\ 3, \; x = 3 \end{cases}$$

$$= \begin{cases} x + 2, \; x \neq 3 \\ 3, \; x = 3 \end{cases}$$

Continuity at $x = 3$ requires

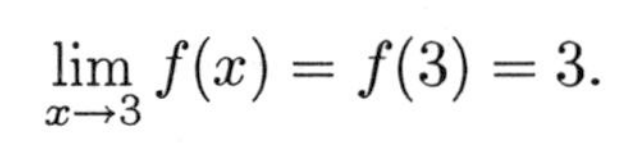

$$\lim_{x \to 3} f(x) = f(3) = 3.$$

Now, $\lim_{x \to 3} f(x) = \lim_{x \to 3} (x + 2) = 5 \neq 3 = f(3)$.

Hence $f(x)$ is *not* continuous at $x = 3$.

(b)

$$f(x) = \frac{x^2 - 4}{x + 2}$$

The function is not defined at $x = -2$ and so cannot be continuous there.

Note:

$$f(x) = \frac{(x-2)(x+2)}{x+2} = x - 2, \; x \neq -2$$

(30%) Q3. In each of the following, find $\dfrac{dy}{dx}$.

(a) $y = \left(x + \dfrac{1}{x^2}\right)^{\sqrt{8}}$

(b) $y = \sqrt{x\sqrt{x\sqrt{x}}}$

(c) $y = \dfrac{\sin x}{\sin(x - \sin x)}$

(d) $x \tan y = y - 1.$

Solution

(a) $y = u^{\sqrt{8}}$, $u = x + x^{-2}$

$$\frac{dy}{dx} = \frac{dy}{du}\frac{du}{dx} = \sqrt{8}u^{\sqrt{8}-1}(1 - 2x^{-3})$$
$$= \sqrt{8}(1 - 2x^{-3})\left(x + \frac{1}{x^2}\right)^{\sqrt{8}-1}$$

(b)

$$y = \sqrt{x\sqrt{x^{\frac{3}{2}}}} = \sqrt{x \cdot x^{\frac{3}{4}}} = \sqrt{x^{\frac{7}{4}}}$$
$$= x^{\frac{7}{8}}$$

Therefore

$$\frac{dy}{dx} = \frac{7}{8}x^{-\frac{1}{8}}$$

(c)

$$\frac{dy}{dx} = \frac{\cos x \sin(x - \sin x) - \sin x \cos(x - \sin x)[1 - \cos x]}{\sin^2(x - \sin x)}$$

(d) $x \tan y = y - 1$

Implicit differentiation:

$$\tan y + x \sec^2 y \frac{dy}{dx} = \frac{dy}{dx}$$

Hence

$$\frac{dy}{dx}(1 - x\sec^2 y) = \tan y$$

Therefore

$$\frac{dy}{dx} = \frac{\tan y}{1 - x\sec^2 y}$$

(20%) Q4. (a) Find y'' if $x^6 + y^6 = 1$.

(b) Find the equation of the tangent to the curve defined by

$$\cos x + \cos y = x^2 + 1$$

at the point $\left(0, \frac{\pi}{2}\right)$.

Solution

(a) Implicit differentiation:

$$6x^5 + 6y^5\frac{dy}{dx} = 0$$

Therefore

$$y' = \frac{dy}{dx} = -\frac{x^5}{y^5} \qquad (*)$$

$$\begin{aligned} y'' &= -\left[\frac{5x^4y^5 - 5y^4y'x^5}{y^{10}}\right] \\ &= -\left[\frac{5x^4y^5 + 5y^4\frac{x^5}{y^5}x^5}{y^{10}}\right] \quad \text{(using (*))} \\ &= -5\left[\frac{x^4}{y^5} - \frac{x^{10}}{y^{11}}\right]. \end{aligned}$$

(b) First get slope of tangent.

Implicit differentiation:

$$-\sin x - \sin y \cdot y' = 2x$$

i.e.

$$y' = \frac{2x + \sin x}{-\sin y}$$

At $\left(0, \frac{\pi}{2}\right)$,

$$y' = \frac{0+0}{-1} = 0$$

Hence, tangent has equation $y = \frac{\pi}{2}$.

SOLUTIONS TO FINAL EXAMINATION #1

(10%) Q1. (i) Evaluate the following limits.

(a) $\lim_{x\to\infty}(\sqrt{x^2+3x-1}-x)$

(b) $\lim_{x\to-\infty}(\sqrt{x^2+3x-1}-x)$

(c) $\lim_{h\to 0}\dfrac{\sin \pi h^2 \tan 3h^2}{h^3}$

(ii) Let

$$f(x)=\begin{cases}1-\sin x, & -2\pi\le x<0,\\ 1, & x=0,\\ x^2-1, & 0<x\le 2.\end{cases}$$

(a) Is $f(x)$ continuous at $x=0$? Explain.

(b) Is $f(x)$ differentiable at $x=0$? Explain.

Solution

(i)(a) Limits as $x\to\pm\infty$ can also be dealt with routinely. Consider the following steps.

Step 1

Try to write the expression in the form

$$\lim_{x\to\infty}\frac{f(x)}{g(x)}$$

– this form is in general, 'easier' to investigate. (The present form $\lim_{x\to\infty}(\sqrt{x^2+3x-1}-x)$ might mislead some into thinking that the limit doesn't exist. However, bear in mind that we are subtracting very large numbers from each other so that there is a possibility that the limit might indeed exist!)

$$\begin{aligned}\sqrt{x^2+3x-1}-x &= (\sqrt{x^2+3x-1}-x)\frac{(\sqrt{x^2+3x-1}+x)}{(\sqrt{x^2+3x-1}+x)}\\ &= \frac{x^2+3x-1-x^2}{\sqrt{x^2+3x-1}+x}\end{aligned}$$

i.e.

$$\sqrt{x^2+3x-1}-x=\frac{3x-1}{\sqrt{x^2+3x-1}+x}$$

Step 2

Divide the top and bottom of the expression by x^n where n is the highest power of x appearing *anywhere* (i.e. top and bottom).

i.e. numerator: highest power appears as x

demoninator: highest power appears as x (i.e. $\sqrt{x^2}$ or x)

Hence

$$\lim_{x\to\infty} \sqrt{x^2+3x-1} - x = \lim_{x\to\infty} \frac{\frac{3x-1}{x}}{\frac{\sqrt{x^2+3x-1}+x}{x}}$$
$$= \lim_{x\to\infty} \frac{3-\frac{1}{x}}{\frac{\sqrt{x^2+3x-1}}{x}+1} \qquad (*)$$

Now be careful!!!

$\sqrt{x^2} = |x| = x$ only if $x > 0$ (which it is in this case since $x \to +\infty$ – see (b) ahead)

Hence

$$\lim_{x\to\infty} \frac{3-\frac{1}{x}}{\frac{\sqrt{x^2+3x-1}}{\sqrt{x^2}}+1} = \lim_{x\to\infty} \frac{3-\frac{1}{x}}{\sqrt{1+\frac{3}{x}-\frac{1}{x^2}}+1}$$

Step 3

Recall

$$\lim_{x\to\infty} \frac{1}{x^r} = 0 \quad (r > 0 \text{ is rational}),$$
$$\lim_{x\to-\infty} \frac{1}{x^r} = 0 \quad (r > 0 \text{ is rational such that } x^r \text{ is defined } \forall\, x).$$

Thus,

$$\lim_{x\to\infty} \frac{3-\frac{1}{x}}{\sqrt{1+\frac{3}{x}-\frac{1}{x^2}}+1} = \frac{3-\lim\limits_{x\to\infty}\frac{1}{x}}{\sqrt{1+\lim\limits_{x\to\infty}\frac{3}{x}-\lim\limits_{x\to\infty}\frac{1}{x^2}}+1}$$
$$= \frac{3-0}{\sqrt{1+0-0}+1} = \frac{3}{2}$$

(Not quite as expected!!!)

(i)(b) The calculation proceeds exactly as in (a) above until we reach (*) in Step 2.

Since $x \to -\infty$ we now have $x < 0$ so that $x = -\sqrt{x^2}$.

Thus, at (*):

$$\lim_{x\to-\infty} \frac{3-\frac{1}{x}}{\frac{\sqrt{x^2+3x-1}}{x}+1} = \lim_{x\to-\infty} \frac{3-\frac{1}{x}}{\frac{\sqrt{x^2+3x-1}}{-\sqrt{x^2}}+1}$$
$$= \lim_{x\to-\infty} \frac{3-\frac{1}{x}}{-\sqrt{1+\frac{3}{x}-\frac{1}{x^2}}+1}$$

This time we get

$$\frac{3-0}{-1+1}$$

i.e.

$$\frac{3}{0}$$

So the limit doesn't exist!

Note that for the expression $\lim_{x\to\pm\infty} \sqrt{x^2+3x-1}+x$, the results would be the other way around, i.e., the limit as $x \to \infty$ doesn't exist but the limit as $x \to -\infty$ is $-\frac{3}{2}$ (verify!).

(i)(c) **Stage 1**

$\left(\frac{0}{0}\right)$

Stage 2

Simplify

$$\frac{\sin \pi h^2 \tan 3h^2}{h^3} = \left(\frac{\sin \pi h^2}{\pi h^2}\right)\left(\frac{\pi}{h}\right)\left(\frac{\sin 3h^2}{\cos 3h^2}\right)$$
$$= \left(\frac{\sin \pi h^2}{\pi h^2}\right)\cdot\frac{\pi}{h}\left(\frac{\sin 3h^2}{3h^2}\right)\cdot 3h^2\left(\frac{1}{\cos 3h^2}\right)$$
$$= \left(\frac{\sin \pi h^2}{\pi h^2}\right)\cdot\pi\left(\frac{\sin 3h^2}{3h^2}\right)\cdot 3h\cdot\left(\frac{1}{\cos 3h^2}\right)$$

Stage 3

$$\lim_{h\to0} \frac{\sin \pi h^2 \tan 3h^2}{h^3}$$
$$= \underbrace{\lim_{h\to0}\left(\frac{\sin \pi h^2}{\pi h^2}\right)}_{1}\pi\underbrace{\lim_{h\to0}\left(\frac{\sin 3h^2}{3h^2}\right)}_{1}\underbrace{\lim_{h\to0} 3h}_{0}\underbrace{\lim_{h\to0}\left(\frac{1}{\cos 3h^2}\right)}_{1}$$
$$= 1\cdot\pi\cdot1\cdot0\cdot1 = 0$$

(ii)(a) $f(x)$ is continuous at $x = 0$ iff

$$\lim_{x \to 0} f(x) = f(0) = 1$$

Now,

$$\lim_{x \to 0^-} f(x) = \lim_{x \to 0^-} (1 - \sin x) = 1$$
$$\lim_{x \to 0+} f(x) = \lim_{x \to 0+} (x^2 - 1) = -1$$

Hence $\lim_{x \to 0} f(x)$ doesn't exist! Thus $f(x)$ is *not* continuous at $x = 0$.

(ii)(b)

$$f'(0) = \lim_{h \to 0} \frac{f(0+h) - f(0)}{h} = \lim_{h \to 0} \frac{f(h) - 1}{h}$$

Hence

$$\lim_{h \to 0+} \frac{f(h) - 1}{h} = \lim_{h \to 0+} \frac{h^2 - 1 - 1}{h}$$
$$= \lim_{h \to 0+} \frac{h^2 - 2}{h}$$

which doesn't exist $\left(\text{i.e. Stage 1 gives } \left(\frac{-2}{0}\right)\right)$.

Hence $f'(0)$ doesn't exist and $f(x)$ is *not* differentiable at $x = 0$.

(15%) Q2. Evaluate the following definite integrals:

(a) $\displaystyle\int_0^1 |x^3 - x - 3x^2 + 3| dx$

(b) $\displaystyle\int_0^1 \frac{2x^3 + 9x^2 + 5}{(x^4 + 6x^3 + 10x + 7)^{\frac{1}{2}}} dx$

(c) $\displaystyle\int_{-1}^1 x^{-\frac{8}{9}} \sin\left(x^{\frac{1}{9}}\right) dx$

(d) $\displaystyle\int_{-a}^a x^3 \cos(x^3) dx, \quad a \in \mathbb{R}.$

Solution

(a)

$$|x^3 - x - 3x^2 + 3| = |(x-1)(x+1)(x-3)|$$

Now,

$$\begin{aligned}
(x-1)(x+1)(x-3) \geq 0 \quad \text{when} \quad -1 \leq x \leq 1 \quad \text{or} \quad x \geq 3 \\
(x-1)(x+1)(x-3) \leq 0 \quad \text{when} \quad x \leq -1 \quad \text{or} \quad 1 \leq x \leq 3
\end{aligned}$$

(See Section 3.8, Part 1)

Hence

$$|x^3 - x - 3x^2 + 3| = \begin{cases} x^3 - x - 3x^2 + 3, & -1 \leq x \leq 1 \quad \text{or} \quad x \geq 3, \\ -(x^3 - x - 3x^2 + 3), & x \leq -1 \quad \text{or} \quad 1 \leq x \leq 3 \end{cases}$$

Thus

$$\begin{aligned}
\int_0^1 |x^3 - x - 3x^2 + 3| dx &= \int_0^1 (x^3 - x - 3x^2 + 3) dx \\
&= \left[\frac{x^4}{4} - \frac{x^2}{2} - x^3 + 3x \right]_0^1 \\
&= \left(\frac{1}{4} - \frac{1}{2} - 1 + 3 \right) \\
&= \frac{7}{4}
\end{aligned}$$

(b) Let $u = x^4 + 6x^3 + 10x + 7$.

Then

$$\begin{aligned}
du &= (4x^3 + 18x^2 + 10) dx \\
&= 2(2x^3 + 9x^2 + 5) dx
\end{aligned}$$

Also,

$$x = 0, \; u = 7; \; x = 1, \; u = 24$$

Hence

$$I = \frac{1}{2} \int_7^{24} u^{-\frac{1}{2}} du = \frac{1}{2} \left[2u^{\frac{1}{2}} \right]_7^{24} = \sqrt{24} - \sqrt{7}$$

(c) Let $u = x^{\frac{1}{9}} \; du = \frac{1}{9} x^{-\frac{8}{9}} dx$

When $x = -1$, $u = -1$

When $x = 1$, $u = 1$

Therefore

$$I = 9\int_{-1}^{1} \sin u du = -9\left[\cos u\right]_{-1}^{1} = -9(\cos(1) - \cos(-1))$$
$$= 0$$

(d) This is of the form $\int_{-a}^{a} f(x)dx$ where $f(x)$ is an odd function, i.e. $f(-x) = -f(x)$. Hence $\int_{-a}^{a} x^3 \cos(x^3)dx = 0$. (See Appendix.)

(15%) Q3. Find the following.

(a) $\displaystyle\int \frac{\cos^3 x \sin x}{\sqrt{1+\cos^4 x}}dx$

(b) $\displaystyle\int \frac{x^{-3} + x^6\sqrt{1+x^5}}{5x^2}dx$

(c) $\displaystyle\int \frac{\cos\left(\frac{1}{t}\right)}{t^2}dt$

(d) $\displaystyle\int \sin^4 t \cos^3 t dt$

(e) $f'(x)$ where $f(x) = \displaystyle\int_{x}^{x^2+1} \tan\left(\sqrt{u^2+1}\right) du$

Solution

(a) Let

$$u = 1 + \cos^4 x,\ du = 4\cos^3 x(-\sin x)dx$$
$$= -4\sin x \cos^3 x dx$$

Therefore

$$I = -\frac{1}{4}\int u^{-\frac{1}{2}}du = -\frac{1}{4}2u^{\frac{1}{2}} + C$$

Therefore

$$I = -\frac{1}{2}(1+\cos^4 x)^{\frac{1}{2}} + C, \quad C \text{ is an arbitrary constant}$$

(b)

$$
\begin{aligned}
I &= \int \left[\frac{1}{5x^5} + \frac{x^4}{5}\sqrt{1+x^5} \right] dx \\
&= \frac{1}{5}\int x^{-5}dx + \frac{1}{25}\int \sqrt{u}du \quad (u = 1 + x^5,\ du = 5x^4dx) \\
&= \frac{1}{5}\frac{x^{-4}}{-4} + \frac{1}{25}\frac{u^{\frac{3}{2}}}{\frac{3}{2}} + C \\
&= -\frac{1}{20x^4} + \frac{2}{75}(1+x^5)^{\frac{3}{2}} + C
\end{aligned}
$$

(c) Let $u = \frac{1}{t}$; $du = -\frac{1}{t^2}dt$

$$
\begin{aligned}
I = -\int \cos u du &= -\sin u + C \\
&= -\sin\left(\frac{1}{t}\right) + C
\end{aligned}
$$

(d) Let $u = \sin t,\ du = \cos t dt$

$$
\sin^4 t \cos^3 t dt = u^4(1-u^2)du \quad (\text{since } \cos^2 t = 1 - \sin^2 t)
$$

Therefore

$$
I = \int (u^4 - u^6)du = \frac{u^5}{5} - \frac{u^7}{7} + C
$$

Therefore

$$
I = \frac{\sin^5 t}{5} - \frac{\sin^7 t}{7} + C
$$

(e) First write the integral in a form suitable for the Fundamental Theorem of Calculus (see Appendix).

i.e. Write

$$
f(x) = \left(\int_x^a + \int_a^{x^2+1} \right) \tan\left(\sqrt{u^2+1}\right) du,
$$

where a is some number in the domain of $g(u) = \tan\sqrt{u^2+1}$.

i.e.

$$f(x) = \underbrace{-\int_a^x \tan(\sqrt{u^2+1})du}_{\text{Apply Fundamental Theorem immediately}} + \underbrace{\int_a^{x^2+1} \tan\sqrt{u^2+1}du}_{\text{Need to use chain rule before applying Fundamental Theorem}}$$

Therefore

$$f'(x) = -\tan\sqrt{x^2+1} + \frac{d}{dx}\int_a^{x^2+1} \tan\sqrt{u^2+1}du$$

(The Fundamental Theorem of Calculus says $\frac{d}{dw}\int_a^w f(u)du = f(w)$. Note the required form, i.e. the *constant* a is at the lower limit of integration and the *variable of differentiation* at the upper limit – see Appendix. Hence, here it is necessary to split the integral and use the chain rule.)

Write

$$\begin{aligned} f'(x) &= -\tan\sqrt{x^2+1} + \frac{d}{dx}\int_a^w \tan\sqrt{u^2+1}du, \ w = x^2+1 \\ &= -\tan\sqrt{x^2+1} + \underbrace{\frac{d}{dw}\int_a^w \tan\sqrt{u^2+1}du \cdot \frac{dw}{dx}}_{\text{Chain Rule}} \end{aligned}$$

Now we can once again apply the Fundamental Theorem of Calculus:

$$\begin{aligned} f'(x) &= -\tan\sqrt{x^2+1} + \tan\sqrt{w^2+1} \cdot 2x \\ &= -\tan\sqrt{x^2+1} + 2x\tan\sqrt{(x^2+1)^2+1} \end{aligned}$$

(20%) Q4. Discuss the graph of $y = f(x) = \frac{7x^2}{x^2-1}$ under the following headings.

(a) Domain of $f(x)$.

(b) x and y-axis intercepts.

(c) Asymptotes.

(d) Intervals of increase or decrease.

(e) Local maximum and minimum values.

(f) Concavity and points of inflection.

Sketch the curve.

Solution

$$y = f(x) = \frac{7x^2}{x^2 - 1}$$

(a) **Domain**: We remove values of x at which $f(x)$ is undefined, i.e. $x:\ x^2 - 1 = 0$, i.e. $x = \pm 1$. Hence, domain of f is $\{x \in \mathbb{R} : x \neq \pm 1\}$.

(b) x-intercept : Let $y = 0$, i.e. $\dfrac{7x^2}{x^2 - 1} = 0$, i.e. $x = 0$

y-intercept : Let $x = 0$, i.e. $y = 0$

(c) **Asymptotes**:

(i) Horizontal: Write $y = 7\left(1 + \dfrac{1}{x^2 - 1}\right)$

Now, $\lim\limits_{x \to \pm\infty} f(x) = 7$

Hence, $y = 7$ is a horizontal asymptote.

(ii) Vertical: Look for values of x at which the denominator becomes zero, i.e. $x = \pm 1$

Hence, $x = \pm 1$ are vertical asymptotes. For each asymptote we have a choice of how $f(x)$ tends to infinity as x approaches ± 1.

For example,

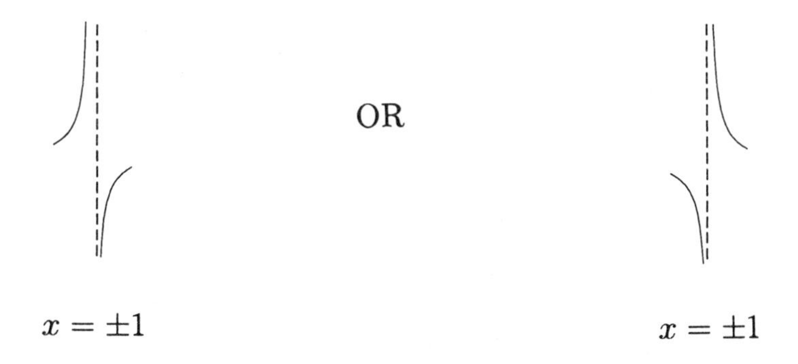

To determine which behaviour is appropriate in this case, we look at the limit as $x \to \pm 1^{\pm}$.

i.e.

$\underline{x = 1}$

$$\left.\begin{aligned} f(x) &= \frac{7x^2}{(x-1)(x+1)} \\ \lim_{x\to 1^+} f(x) &= \frac{7}{(\text{small positive no.})(2)} = +\infty \\ \lim_{x\to 1^-} f(x) &= \frac{7}{(\text{small negative no.})(2)} = -\infty \end{aligned}\right\} \text{i.e.}$$

$x = 1$

$\underline{x = -1}$

$$\left.\begin{aligned} f(x) &= \frac{7x^2}{(x-1)(x+1)} \\ \lim_{x\to -1^+} f(x) &= \frac{7}{(-2)(\text{small positive no.})} = -\infty \\ \lim_{x\to -1^-} f(x) &= \frac{7}{(-2)(\text{small negative no.})} = +\infty \end{aligned}\right\} \text{i.e.}$$

$x = -1$

(d) Intervals of increase or decrease

$$f'(x) = \frac{14x(x^2-1) - 2x \cdot 7x^2}{(x^2-1)^2} = -\frac{14x}{(x^2-1)^2}, \quad x \neq \pm 1.$$

f increases when $f' > 0$ i.e. $x \in (-\infty, -1) \cup (-1, 0)$
f decreases when $f' < 0$ i.e. $x \in (0, 1) \cup (1, \infty)$

(See Section 3.8, Part 1.)

(e) Local maximum and local minimum values

$$f'(x) = 0 \Leftrightarrow -\frac{14x}{(x^2-1)^2} = 0 \Leftrightarrow x = 0$$

Hence, $(0, 0)$ is a local maximum (using (d)). No local minima.

(f) **Concavity and points of inflection**

$$\begin{aligned} f''(x) &= -14\left\{\frac{1\cdot(x^2-1)^2 - 2(x^2-1)(2x)x}{(x^2-1)^4}\right\} \\ &= -\frac{14}{(x^2-1)^4}\left\{(x^2-1)((x^2-1)-4x^2)\right\} \\ &= -\frac{14(-3x^2-1)}{(x^2-1)^3} = \frac{14(3x^2+1)}{(x^2-1)^3}, \quad x \neq \pm 1. \end{aligned}$$

To determine when this expression is positive or negative consider the following:

$$\begin{aligned} 3x^2+1 &> 0 \;\forall\, x \\ (x^2-1)^3 &> 0 \Leftrightarrow x^2-1>0 \quad \text{i.e.} \quad x \in (-\infty,-1)\cup(1,\infty) \end{aligned}$$

Hence

$$\begin{aligned} f'' &> 0 \quad (f \quad \text{is concave up) for} \quad x \in (-\infty,-1)\cup(1,\infty) \\ f'' &< 0 \quad (f \quad \text{is concave down) for} \quad x \in (-1,1) \end{aligned}$$

No points of inflection ($x = \pm 1$ are asymptotes!)

Sketch

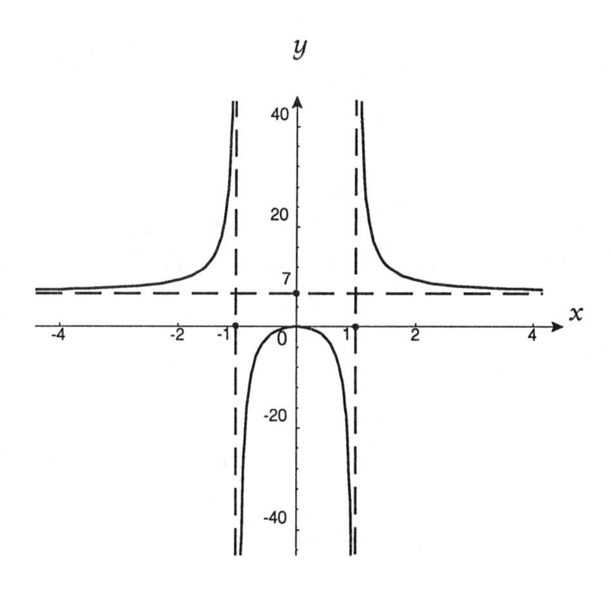

(10%) Q5. An open-top cylindrical can is made to hold 6 litres of oil. Find the dimensions that will minimize the cost of the metal.

Solution

h r

Volume $= V = \pi r^2 h = b$

Surface Area $= S = \pi r^2 + 2\pi r h$

Now,

$$S(r) = \pi r^2 + 2\pi r\left(\frac{b}{\pi r^2}\right) = \pi r^2 + \frac{2b}{r}$$

$$S'(r) = 2\pi r - \frac{2b}{r^2} = 0 \Leftrightarrow 2\pi r^3 = 2b$$

i.e.

$$r = \left(\frac{b}{\pi}\right)^{\frac{1}{3}}.$$

This gives an absolute minimum since

$$S'(r) < 0, \quad 0 < r < \left(\frac{b}{\pi}\right)^{\frac{1}{3}},$$

$$S'(r) > 0, \quad r > \left(\frac{b}{\pi}\right)^{\frac{1}{3}}$$

(See Section 3.8, Part 1).

The dimensions of the can are thus,

$$r = \left(\frac{b}{\pi}\right)^{\frac{1}{3}},$$

$$h = \frac{b}{\pi r^2} = \frac{b}{\left[\pi\left(\frac{b}{\pi}\right)^{\frac{2}{3}}\right]} = \left(\frac{b}{\pi}\right)^{\frac{1}{3}}.$$

(10%) Q6. Find the limit if it exists. If the limit does not exist, explain why.

(a) $\lim_{\theta \to 0} \left(\theta^2 \cot\theta + 5\right)$

(b) $\lim_{x \to 0} \left(\frac{1 - \sin x}{x^3}\right)$

(c) $\lim_{x \to 1+} \frac{1 - \sqrt{x}}{1 - x}$

(d) $\lim_{x \to -\infty} \frac{x-1}{x^2}$

Solution

(a)

$$\lim_{\theta \to 0} \left(\theta^2 \frac{\cos\theta}{\sin\theta} + 5 \right) = \lim_{\theta \to 0} \left(\theta \cdot \frac{\theta}{\sin\theta} \cdot \cos\theta + 5 \right)$$

$$= \underbrace{\lim_{\theta \to 0} \theta}_{=0} \cdot \underbrace{\lim_{\theta \to 0} \frac{\theta}{\sin\theta}}_{=1} \cdot \underbrace{\lim_{\theta \to 0} \cos\theta}_{=1} + \underbrace{\lim_{\theta \to 0} 5}_{=5}$$

$$= 5$$

(b) **Stage 1**

$\left(\frac{1}{0}\right)$ – limit does not exist.

i.e. the numerator tends towards 1 while the denominator gets smaller. Hence the expression grows to infinity.

(c) **Stage 1**

$\left(\frac{0}{0}\right)$

Stage 2

Simplify the expression:

$$\frac{1-\sqrt{x}}{1-x} \cdot \frac{1+\sqrt{x}}{1+\sqrt{x}} = \frac{1-x}{(1-x)(1+\sqrt{x})}$$

Hence,

$$\lim_{x \to 1^+} \frac{1-\sqrt{x}}{1-x} = \lim_{x \to 1^+} \frac{(1-x)}{(1-x)(1+\sqrt{x})}$$

$$= \lim_{x \to 1^+} \frac{1}{1+\sqrt{x}}$$

Stage 3

$$\lim_{x \to 1^+} \frac{1}{1+\sqrt{x}} = \frac{1}{2}$$

(d) Divide numerator and denominator by x^2 (highest power)

$$\lim_{x\to-\infty} \frac{x-1}{x^2} = \lim_{x\to-\infty} \frac{\frac{1}{x} - \frac{1}{x^2}}{1} = 0$$

(10%) Q7. Find the area bounded by the graphs of $y = x^3$ and $y = x$.

Solution

Curves intersect when $x^3 - x = 0$, i.e. $x(x-1)(x+1) = 0$, i.e. $x = 0, \pm 1$.

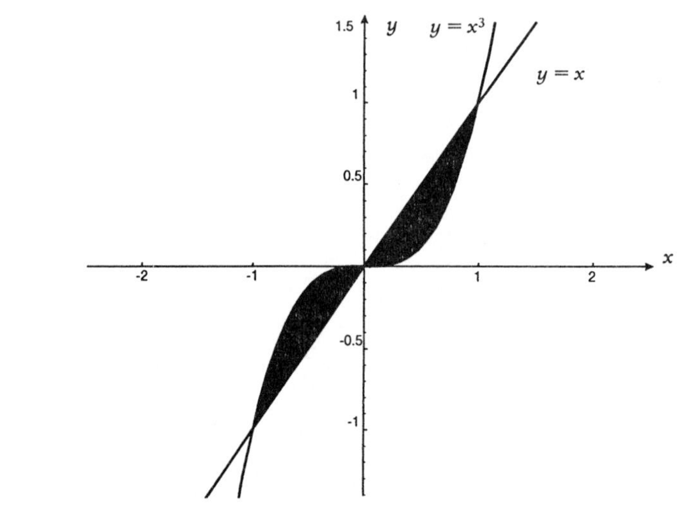

$$\begin{aligned} A = \int_{-1}^{1} |x^3 - x| dx &= \int_{-1}^{0} (x^3 - x) dx + \int_{0}^{1} (x - x^3) dx \\ &= 2\int_{0}^{1} (x - x^3) dx \\ &= 2\left[\frac{x^2}{2} - \frac{x^4}{4}\right]_0^1 = 2\left(\frac{1}{4}\right) = \frac{1}{2} \end{aligned}$$

(10%) Q8. The height h of a triangle is increasing at a rate of 5cm/min while the area of the triangle is increasing at a rate of 8cm^2/min. At what rate is the base of the triangle changing when the height is 50cm and the area 500cm^2?

Solution

Area $= \frac{1}{2}bh$ where b is base of triangle.

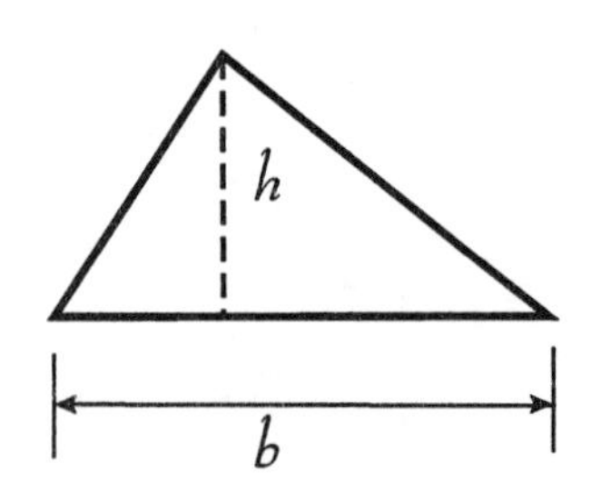

We want: $\frac{db}{dt}$ when $h = 50,\ A = 500$.

We know: $\frac{dh}{dt}, \frac{dA}{dt}$

Relate what you *know* to what you *want*.

$$\frac{dA}{dt} = \frac{1}{2}\left(b\frac{dh}{dt} + h\frac{db}{dt}\right)$$

Hence,

$$\frac{db}{dt} = \frac{1}{h}\left[2\frac{dA}{dt} - b\frac{dh}{dt}\right] = \frac{1}{50}(2(8) - b(50)) \qquad (*)$$

To get b at this time, note that

$$\text{Area } = 500 = \frac{1}{2}b \cdot 50 \quad \text{i.e. } b = 20$$

Hence, from (*), $\frac{db}{dt} = \frac{16 - 100}{50} = -\frac{84}{50}$cm/min.

SOLUTIONS TO FINAL EXAMINATION #2

(30%) 1. Evaluate when possible:

(a) $\lim_{x\to 0} \frac{x - x\cos 3x}{x + x\cos 3x}$ **OR*** $\lim_{x\to 0} \frac{\sin x - x - x^3}{x^2}$

(b) $\lim_{x\to 2} \left(\frac{1}{x-2} - \frac{2}{x^2-4}\right)$ **OR*** $\lim_{x\to\infty} \frac{(\ln x)^2}{x}$

(c) $\lim_{x\to -\infty} \frac{8x^3 + 2x^2 - 1}{(x^9 - 4x + 11)^{\frac{1}{3}}}$ **OR*** $\lim_{x\to 0+} (3x+1)^{\frac{2}{x}}$

(d) $\int_0^2 |x^3 - 3x^2 + 2x| dx$ **OR*** $\int_1^8 e^{\sqrt[3]{x}} x^{-\frac{2}{3}} dx$

(e) $\lim_{n\to\infty} \sum_{i=1}^{n} \frac{8}{n} \left[3\left(2 + \frac{2i}{n}\right)^2 + 2\left(2 + \frac{2i}{n}\right)\right]$.

Solution

(a) **Stage 1**

$\left(\frac{0}{0}\right)$

Stage 2

Simplify

$$\lim_{x\to 0} \frac{x - x\cos 3x}{x + x\cos 3x} = \lim_{x\to 0} \frac{1 - \cos 3x}{1 + \cos 3x}$$

Stage 3

$$\lim_{x\to 0} \frac{1 - \cos 3x}{1 + \cos 3x} = \frac{0}{2} = 0$$

OR*

Use l'Hôpital's rule.

$$\begin{aligned}
&\lim_{x\to 0} \frac{\sin x - x - x^3}{x^2} && \left(\text{indeterminate of form } \left(\frac{0}{0}\right)\right) \\
\underset{\text{l'Hôpital}}{=}\ &\lim_{x\to 0} \frac{\cos x - 1 - 3x^2}{2x} && \left(\text{still of the form } \left(\frac{0}{0}\right)\right) \\
\underset{\text{l'Hôpital}}{=}\ &\lim_{x\to 0} \frac{-\sin x - 6x}{2} && \left(\text{now of the form } \left(\frac{0}{2}\right)\right) \\
=\ &0
\end{aligned}$$

(b) **Stage 1**

$\left(\frac{1}{0} - \frac{2}{0}: \quad \text{meaningless!}\right)$

Stage 2

Simplify:

$$\frac{1}{x-2} - \frac{2}{(x^2-4)} = \frac{1}{x-2} - \frac{2}{(x-2)(x+2)} = \frac{x}{(x-2)(x+2)}$$

Stage 3

$$\lim_{x \to 2} \frac{x}{(x-2)(x+2)} = \frac{2}{0}$$

i.e. limit doesn't exist!

OR*

$$\begin{aligned}
&\lim_{x\to\infty} \frac{(\ln x)^2}{x} \qquad \left(\text{indeterminate of form } \left(\frac{\infty}{\infty}\right)\right) \\
\underset{\text{l'Hôpital}}{=} &\lim_{x\to\infty} \frac{2 \cdot \ln x \cdot \frac{1}{x}}{1} \\
= &\lim_{x\to\infty} \frac{2\ln x}{x} \qquad \left(\text{still of the form } \left(\frac{\infty}{\infty}\right)\right) \\
\underset{\text{l'Hôpital}}{=} &\lim_{x\to\infty} \frac{2}{x} \\
= &\ 0
\end{aligned}$$

(c) **Step 1** Expression is already in the form $\lim\limits_{x\to-\infty} \dfrac{f(x)}{g(x)}$.

Step 2 Divide top and bottom by highest power: x^3.

$$\frac{8 + \frac{2}{x} - \frac{1}{x^3}}{\left(1 - \frac{4}{x^8} + \frac{11}{x^9}\right)^{\frac{1}{3}}} \qquad \left(\text{writing} \quad x^3 = (x^9)^{\frac{1}{3}}\right)$$

Step 3

$$\lim_{x\to-\infty} \frac{8 + \frac{2}{x} - \frac{1}{x^3}}{\left(1 - \frac{4}{x^8} + \frac{11}{x^9}\right)^{\frac{1}{3}}} = \frac{8}{1} = 8$$

OR*

$$\lim_{x\to 0^+}(3x+1)^{\frac{2}{x}} \qquad \text{(indeterminate of form } 1^\infty\text{)}$$

Let

$$u = (3x+1)^{\frac{2}{x}}$$

Then

$$\ln u = \frac{2}{x}\ln(3x+1)$$

Thus

$$\begin{aligned}\lim_{x\to 0^+}\ln u &= \lim_{x\to 0^+}\frac{2\ln(3x+1)}{x} \qquad \left(\text{form}\left(\frac{0}{0}\right)\right)\\ &\underset{\text{l'Hôpital}}{=} \lim_{x\to 0^+}\frac{2\cdot 3}{(3x+1)}\\ &= 6\end{aligned}$$

(d)

$$\begin{aligned}|x^3-3x^2+2x| &= |x(x-1)(x-2)|\\ &= \begin{cases} x(x-1)(x-2), & x \in (0,1) \cup (2,\infty)\\ -x(x-1)(x-2), & x \in (-\infty,0) \cup (1,2)\end{cases}\end{aligned}$$

(See Section 3.8, Part 1.)

Hence

$$\begin{aligned}&\int_0^2 |x^3-3x^2+2x|dx\\ &= \int_0^1 (x^3-3x^2+2x)dx - \int_1^2 (x^3-3x^2+2x)dx\\ &= \left[\frac{x^4}{4}-x^3+x^2\right]_0^1 - \left[\frac{x^4}{4}-x^3+x^2\right]_1^2\\ &= \left(\frac{1}{4}-1+1\right) - \left(\frac{16}{4}-8+4-\frac{1}{4}+1-1\right)\\ &= \frac{1}{4}-\left(-\frac{1}{4}\right)=\frac{1}{2}\end{aligned}$$

OR*

Let $u = x^{\frac{1}{3}}$; $du = \frac{1}{3}x^{-\frac{2}{3}}dx$

$$x = 1,\ u = 1;\ x = 8,\ u = 2$$

$$I = 3\int_1^2 e^u du = 3\left[e^u\right]_1^2$$
$$= 3(e^2 - e)$$

(e)

$$\lim_{n\to\infty}\sum_{i=1}^{n}\frac{8}{n}\left[3\left(4+\frac{8i}{n}+\frac{4i^2}{n^2}\right)+4+\frac{4i}{n}\right]$$
$$= \lim_{n\to\infty}\frac{8}{n}\sum_{i=1}^{n}\left(16+\frac{28i}{n}+\frac{12i^2}{n^2}\right)$$
$$= \lim_{n\to\infty}\left\{\frac{8\cdot 16}{n}\sum_{i=1}^{n}1+\frac{8\cdot 28}{n^2}\sum_{i=1}^{n}i+\frac{8\cdot 12}{n^3}\sum_{i=1}^{n}i^2\right\}$$

Now,

$$\sum_{i=1}^{n}1 = n,\ \sum_{i=1}^{n}i = \frac{n(n+1)}{2},\ \sum_{i=1}^{n}i^2 = \frac{n(n+1)(2n+1)}{6}$$

Hence the above limit becomes:

$$\lim_{n\to\infty}\left\{\frac{8\cdot 16}{n}\cdot n+\frac{8\cdot 28}{n^2}\cdot\frac{n(n+1)}{2}+\frac{8\cdot 12}{n^3}\frac{n(n+1)(2n+1)}{6}\right\}$$
$$= \lim_{n\to\infty}\left\{128+112\frac{(n^2+n)}{n^2}+16\frac{(2n^3+3n^2+n)}{n^3}\right\}$$
$$= 128+112\lim_{n\to\infty}\left(1+\frac{1}{n}\right)+16\lim_{n\to\infty}\left(2+\frac{3}{n}+\frac{1}{n^2}\right)$$
$$= 128+112+32 \quad \left(\text{since} \quad \lim_{n\to\infty}\frac{1}{n} = \lim_{n\to\infty}\frac{1}{n^2} = 0\right)$$
$$= 272$$

(20%) Q2. Find: (a) $f'(x)$ if $f(x) = \left[10 + \left(1+\sqrt{3+x}\right)^{\frac{3}{2}}\right]^{\frac{3}{2}}$

OR* $f(x) = e^{\sqrt{x}}\cos(x^2)$

(b) $g'(x)$ if $g(x) = \cos\left[\tan\left(\sin\left(x^2+1\right)\right)\right]^{\frac{1}{2}}$

OR* $g(x) = \ln[\ln(x^4)]$

(c) $\dfrac{dy}{dx}$ if $x^2y = \cos(xy)$

OR* $e^{x\sin y} + x^3y^2 = \ln(x^2+y^2)$

(d) $y'(x)$ if $y(x) = \int_{\sqrt{x}}^{1} \sqrt{2+\tan^2(t^2)}dt$

(Do not simplify your answers.)

Solution

(a) Let $f(x) = y = u^{\frac{3}{2}}$ where $u = 10 + \left(1+\sqrt{3+x}\right)^{\frac{3}{2}}$

$$f'(x) = \frac{dy}{dx} = \frac{dy}{du}\frac{du}{dx} = \frac{3}{2}u^{\frac{1}{2}}\left[\frac{3}{2}\left(1+\sqrt{3+x}\right)^{\frac{1}{2}}\right]\frac{1}{2}(3+x)^{-\frac{1}{2}}$$
$$= \frac{9}{8}\left[10 + \left(1+\sqrt{3+x}\right)^{\frac{3}{2}}\right]\left(1+\sqrt{3+x}\right)^{\frac{1}{2}}(3+x)^{-\frac{1}{2}}$$

OR*

$$f'(x) = \cos(x^2)\frac{d}{dx}\left(e^{\sqrt{x}}\right) + e^{\sqrt{x}}\frac{d}{dx}(\cos(x^2))$$
$$= \cos(x^2)\cdot\frac{d}{du}e^u\cdot\frac{du}{dx} + e^{\sqrt{x}}\frac{d}{dw}\cos w\frac{dw}{dx}$$

(using chain rule with $u = \sqrt{x},\ w = x^2$)

$$= \cos(x^2)\cdot e^u\cdot\frac{1}{2\sqrt{x}} + e^{\sqrt{x}}(-\sin w)\cdot 2x$$
$$= e^{\sqrt{x}}\left[\frac{\cos(x^2)}{2\sqrt{x}} - 2x\sin(x^2)\right]$$

(b) Let $g(x) = y = \cos u$ where $u = \left[\tan\left(\sin(x^2+1)\right)\right]^{\frac{1}{2}}$

$$g'(x) = \frac{dy}{dx} = \frac{dy}{du}\frac{du}{dx} = -\sin u\frac{du}{dx}$$

To get $\dfrac{du}{dx}$, write $u = w^{\frac{1}{2}},\ w = \tan(\sin(x^2+1))$.

$$\begin{aligned}\frac{du}{dx} = \frac{du}{dw}\frac{dw}{dx} &= \frac{1}{2}w^{-\frac{1}{2}} \cdot \sec^2(\sin(x^2+1))\cos(x^2+1)\cdot 2x \\ &= \frac{2x\sec^2(\sin(x^2+1))\cos(x^2+1)}{2\sqrt{\tan(\sin(x^2+1))}}\end{aligned}$$

Hence

$$g'(x) = \frac{-\sin\left[\tan(\sin(x^2+1))\right]^{\frac{1}{2}} \cdot 2x\sec^2(\sin(x^2+1))\cos(x^2+1)}{2\sqrt{\tan(\sin(x^2+1))}}$$

OR*

Write $g(x) = y = \ln u,\ u = \ln x^4 = 4\ln x$

$$\begin{aligned}g'(x) = \frac{dy}{dx} &= \frac{dy}{du}\frac{du}{dx} \qquad \text{(chain rule)} \\ &= \frac{1}{u}\cdot\frac{4}{x} \\ &= \frac{4}{x\ln x^4}\end{aligned}$$

(c) Implicit differentiation:

$$2xy + x^2y' = -\sin(xy)\left[1\cdot y + xy'\right].$$

i.e.

$$y' = \frac{dy}{dx} = \frac{-(y\sin(xy)+2xy)}{x^2 + x\sin(xy)}$$

OR*

Implicit differentiation

$$(1\cdot\sin y + xy'\cos y)e^{x\sin y} + 3x^2y^2 + 2x^3yy' = (2x+2yy')\ln(x^2+y^2)$$

i.e.

$$y'\left[x\cos y e^{x\sin y} + 2x^3y - 2y\ln(x^2+y^2)\right] = -\sin y e^{x\sin y} - 3x^2y^2 + 2x\ln(x^2+y^2)$$

i.e.

$$y' = \frac{-\sin y e^{x\sin y} - 3x^2y^2 + 2x\ln(x^2+y^2)}{x\cos y e^{x\sin y} + 2x^3y - 2y\ln(x^2+y^2)}$$

(d) Write

$$
\begin{aligned}
y(x) &= -\int_1^{\sqrt{x}} \sqrt{2+\tan^2(t^2)}dt \\
&= -\int_1^{u} \sqrt{2+\tan^2(t^2)}dt, \ u = \sqrt{x} \\
\frac{dy}{dx} = y'(x) &= -\frac{d}{du}\int_1^{u} \sqrt{2+\tan^2(t^2)}dt \cdot \frac{du}{dx}
\end{aligned}
$$

i.e.

$$
y'(x) = -\sqrt{2+\tan^2(u^2)} \cdot \frac{1}{2\sqrt{x}}
$$

(using Fundamental Theorem of Calculus)

i.e.

$$
y'(x) = -\frac{\sqrt{2+\tan^2 x}}{2\sqrt{x}}
$$

(25%) Q3. Find: (a) $\displaystyle\int x^3 \sin(3x^4+10)dx$ **OR*** $\displaystyle\int_{e^2}^{e^3} \frac{dx}{x\ln x}$

(b) $\displaystyle\int \sec^2 x \tan^3 x dx$ **OR*** $\displaystyle\int_0^{\frac{\pi}{2}} \frac{\cos x}{1+\sin x}dx$

(c) $\displaystyle\int_{-\pi}^{\pi} \frac{t^4 \sin t}{1+t^8}dt$ **OR*** $\displaystyle\int_0^1 \frac{e^{\sin^{-1} x}}{\sqrt{1-x^2}}dx$

(d) $\displaystyle\int \frac{\sqrt{1+x^{-5}}}{x^6}dx$ **OR*** $\displaystyle\int \frac{e^x}{e^x+2}dx$

(e) $\displaystyle\int \sin x\,[\sin(\cos x)]\,dx$ **OR*** $\displaystyle\int \frac{\sqrt{\arctan x}}{1+x^2}dx$

Solution

(a) Let $I = \displaystyle\int x^3 \sin(3x^4+10)dx.$

Let $u = 3x^4+10, \ du = 12x^3dx.$

Then

$$
\begin{aligned}
I = \frac{1}{12}\int \sin u du &= -\frac{1}{12}\cos u + C \\
&= -\frac{1}{12}\cos(3x^4+10) + C
\end{aligned}
$$

OR*

Let

$$u = \ln x, \quad du = \frac{1}{x}dx,$$
$$x = e^2, \qquad u = \ln e^2 = 2; \quad x = e^3, \quad u = \ln e^3 = 3.$$

$$I = \int_2^3 \frac{du}{u} = \ln u]_2^3 = \ln 3 - \ln 2$$
$$= \ln \frac{3}{2}$$

(b)

$$\int \sec^2 x \tan^3 x dx$$

Let $u = \tan x, \ du = \sec^2 x dx.$

Therefore

$$I = \int u^3 du = \frac{u^4}{4} + C = \frac{\tan^3 x}{4} + C$$

OR*

Let

$$u = 1 + \sin x, \ du = \cos x dx,$$
$$x = 0, \ u = 1; \ x = \frac{\pi}{2}, \ u = 2.$$

$$I = \int_1^2 \frac{du}{u} = [\ln u]_1^2 = \ln 2$$

(c)

$$\int_{-\pi}^{\pi} \frac{t^4 \sin t}{1 + t^8} dt = 0$$

i.e. Integral is of the form $\int_{-a}^{a} f(t)dt$ where $f(t)$ is *odd*, i.e. $f(-t) = -f(t)$.

OR*

Let $u = \sin^{-1} x, \ du = \dfrac{dx}{\sqrt{1 - x^2}}.$

When: $x=0,\ u=0;\ x=1,\ u=\frac{\pi}{2}$.

Hence

$$
\begin{aligned}
I &= \int_0^{\frac{\pi}{2}} e^u du \\
&= [e^u]_0^{\frac{\pi}{2}} \\
&= e^{\frac{\pi}{2}} - 1
\end{aligned}
$$

(d)

$$
I = \int \frac{\sqrt{1+x^{-5}}}{x^6} dx \qquad \text{Let} \quad u = 1 + x^{-5}
$$

$$
du = -5x^{-6} dx
$$

$$
\begin{aligned}
I &= -\frac{1}{5} \int u^{\frac{1}{2}} du = -\frac{1}{2} \cdot \frac{2}{3} u^{\frac{3}{2}} + C \\
&= -\frac{1}{3} \left(1 + x^{-5}\right)^{\frac{3}{2}} + C
\end{aligned}
$$

OR*

Let $u = e^x + 2;\ du = e^x dx$

$$
\begin{aligned}
I = \int \frac{du}{u} &= \ln|u| + C \\
&= \ln|e^x + 2| + C \\
&= \ln(e^x + 2) + C \quad (\text{since } e^x > 0)
\end{aligned}
$$

(e)

$$
I = \int \sin x \left[\sin(\cos x)\right] dx.
$$

Let $u = \cos x;\ du = -\sin x dx$.

Therefore

$$
\begin{aligned}
I = -\int \sin u du &= \cos u + C \\
&= \cos(\cos x) + C
\end{aligned}
$$

OR*

Let

$$u = \arctan x = \tan^{-1} x$$
$$du = \frac{dx}{1+x^2}$$

$$I = \int \sqrt{u}\, du$$
$$= \frac{2}{3}u^{\frac{3}{2}} + C$$
$$= \frac{2}{3}(\arctan x)^{\frac{3}{2}} + C$$

(10%) Q4. Sketch the graph of $y = f(x) = \dfrac{x}{x-1}$ **OR*** $y = f(x) = \ln(\cos x)$ identifying any critical points, asymptotes, x and y-intercepts, points of inflection and local maxima and minima.

Solution

$$y = f(x) = \frac{x}{x-1} = 1 + \frac{1}{x-1} \qquad (*)$$

$$f'(x) = -\frac{1}{(x-1)^2}; \quad f''(x) = \frac{2}{(x-1)^3}$$

(a) x-intercept at $y = 0$, i.e. $(0, 0)$

y-intercept at $x = 0$, i.e. $(0, 0)$

(b) Horizontal asymptote: $y = 1$ (from (*))

Vertical asymptote: $x = 1$ (from (*))

Also

$$\left.\begin{aligned} \lim_{x\to 1^+} f(x) &= \frac{1}{\text{(small positive no.)}} = +\infty \\ \lim_{x\to 1^-} f(x) &= \frac{1}{\text{(small negative no.)}} = -\infty \end{aligned}\right\} \text{i.e.}$$

$x = 1$

(c) $f'(x) \neq 0$ Therefore no critical points; no local extrema.

(d)

$$f'' > 0 \Leftrightarrow x > 1 \quad \text{i.e. } f \text{ is concave up when } x > 1$$
$$f'' < 0 \Leftrightarrow x < 1 \quad \text{i.e. } f \text{ is concave down when } x < 1$$

No points of inflection ($x = 1$ is a vertical asymptote!)

Sketch

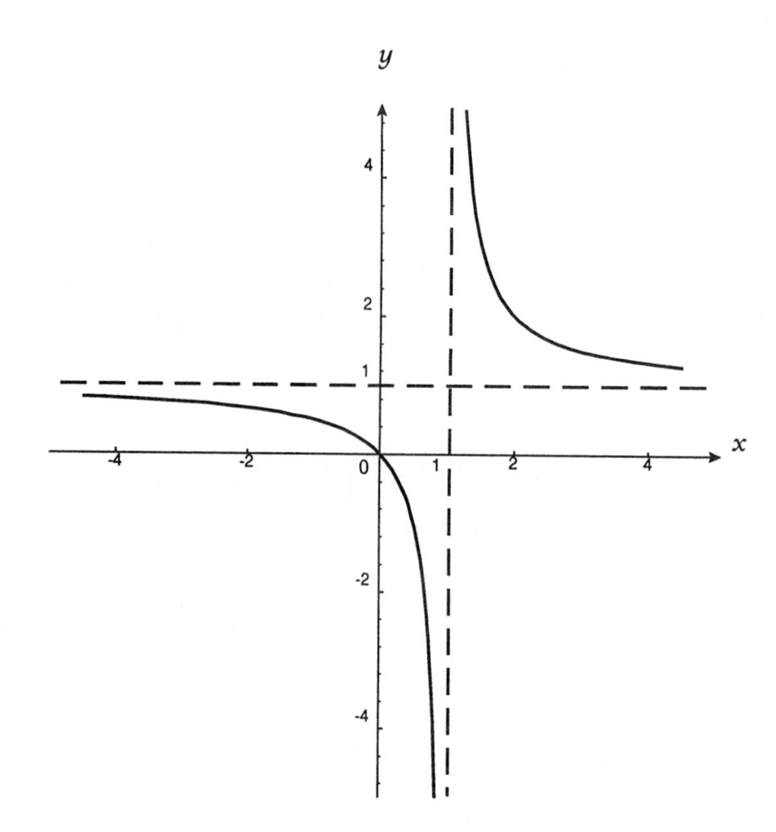

OR*

$$y = f(x) = \ln(\cos x)$$
$$f'(x) = -\frac{\sin x}{\cos x} = -\tan x$$
$$f''(x) = -\sec^2 x$$

(a) f defined only when $\cos x > 0$

i.e. Domain $= \left\{x : 2n\pi - \frac{\pi}{2} < x < 2n\pi + \frac{\pi}{2},\ n = 0, \pm 1, \ldots\right\}$

(b) x-intercept when

$$\begin{aligned} \ln(\cos x) &= 0 \\ &\Leftrightarrow \cos x = 1 \\ &\Leftrightarrow x = 2n\pi \end{aligned}$$

y-intercept at $x = 0$ i.e. $f(0) = 0$.

(c) $f(-x) = f(x)$, so curve is symmetric about y-axis. Further, $f(x + 2\pi) = f(x)$, so f has period 2π.

Consider only $x \in \left(-\frac{\pi}{2}, \frac{\pi}{2}\right)$.

(d) $\lim_{x \to \frac{\pi}{2}^-} \ln(\cos x) = -\infty$; $\lim_{x \to -\frac{\pi}{2}^+} \ln(\cos x) = -\infty$

So $x = \pm\frac{\pi}{2}$ are vertical asymptotes. No horizontal asymptotes.

(e) $f'(x) = -\tan x > 0 \Leftrightarrow -\frac{\pi}{2} < x < 0$

Hence f is increasing on $\left[-\frac{\pi}{2}, 0\right)$ and decreasing on $\left[0, \frac{\pi}{2}\right)$.

Consequently, $f(0) = 0$ is a local maximum.

(f) $f''(x) = -\sec^2 x < 0$ so f is concave down on $\left(-\frac{\pi}{2}, \frac{\pi}{2}\right)$. No inflection points.

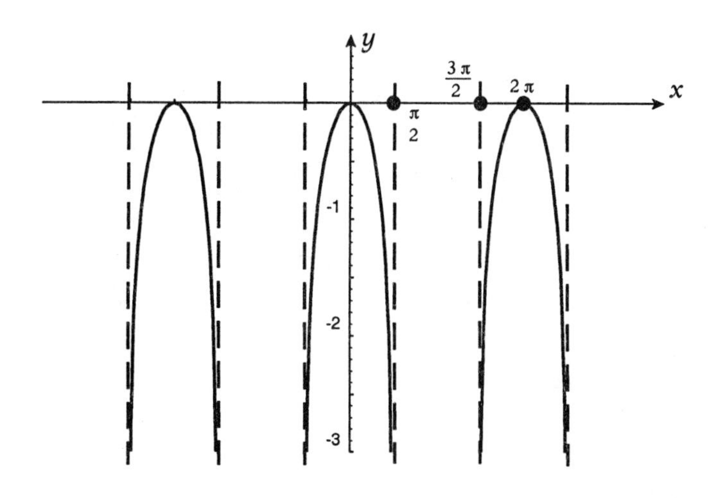

(8%) Q5. Find the total area bounded by the curves $y = x - 2$ and $y^2 - 2x - 4 = 0$.

OR*

Find the Taylor polynomial of degree three, $T_3(x)$, for the function $f(x) = e^x \sin x$ about the point $c = 0$.

Solution

Curves intersect when

$$
\begin{aligned}
(x-2)^2 &= 2x + 4 \\
\text{i.e.} \quad x^2 - 4x + 4 - 2x - 4 &= 0 \\
\text{i.e.} \quad x^2 - 6x &= 0 \\
\text{i.e.} \quad x(x-6) &= 0 \\
\text{i.e.} \quad x &= 0, 6
\end{aligned}
$$

Sketch:

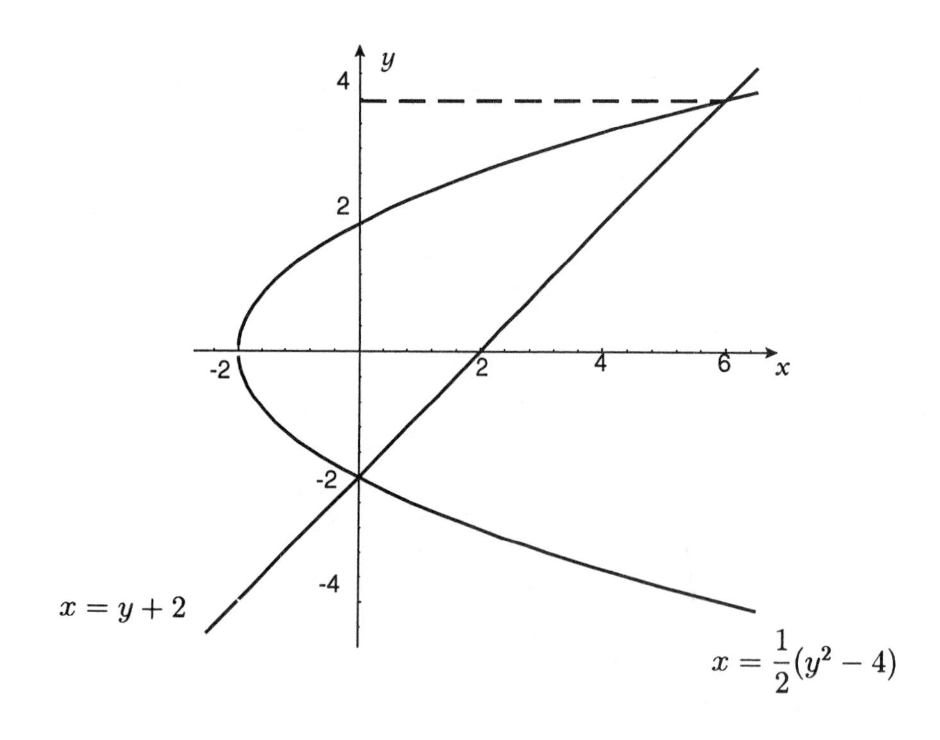

$$\text{Area} = \int_{-2}^{4} \left\{ (y+2) - \frac{1}{2}(y^2 - 4) \right\} dy$$

(easier to describe area with respect to y-variable)

$$= -\int_{-2}^{4} \left(\frac{y^2}{2} - y - 4 \right) dy = -\left[\frac{y^3}{6} - \frac{y^2}{2} - 4y \right]_{-2}^{4}$$

$$= -\left[\left(\frac{64}{6} - 8 - 16 \right) - \left(-\frac{8}{6} - \frac{4}{2} + 8 \right) \right]$$

$$= -\left[\frac{64}{6} - 32 + \frac{20}{6} \right] = -[14 - 32] = 18$$

OR*

$$\begin{aligned} f(x) &= e^x \sin x & f(0) &= 0 \\ f'(x) &= e^x(\sin x + \cos x) & f'(0) &= 1 \\ f''(x) &= 2e^x \cos x & f''(0) &= 2 \\ f^{(3)}(x) &= 2e^x(\cos x - \sin x) & f^{(3)}(0) &= 2 \end{aligned}$$

$$T_3(x) = \sum_{n=0}^{3} \frac{f^{(n)}(0)}{n!} x^n = x + x^2 + \frac{x^3}{3}$$

(7%) Q6. Find the ratio of the height (h) to the radius (r) of the minimum surface-area cone of constant volume b.

(Hint: Surface area S of cone is given by $S^2 = \pi^2(r^4 + r^2h^2)$ and $b = \frac{1}{3}\pi r^2 h$.)

Solution

$$S^2 = \pi^2(r^4 + r^2h^2)$$

$$b = \frac{1}{3}\pi r^2 h \quad \text{therefore} \quad r^2 = \frac{3b}{\pi h}$$

Therefore

$$S^2 = \pi^2 \left(\frac{9b^2}{\pi^2 h^2} + \frac{3hb}{\pi} \right)$$

$$= \frac{9b^2}{h^2} + 3\pi bh$$

Minimize S^2 (and therefore minimize S – it's easier!) with respect to height h.

$$\begin{aligned}\frac{d}{dh}(S^2) &= -\frac{18b^2}{h^3} + 3\pi b \\ &= 0 \\ \Leftrightarrow h^3 &= \frac{6b}{\pi} = 2hr^2 \qquad (*)\end{aligned}$$

i.e.

$$\left(\frac{h^2}{r^2}\right) = 2 \quad \text{i.e.} \quad \left(\frac{h}{r}\right) = \sqrt{2}$$

Note:

The value of h from (*) gives a true minimum since

$$\begin{aligned}\frac{d}{dh}(S^2) &< 0, \quad 0 < h < \left(\frac{6b}{\pi}\right)^{\frac{1}{3}}, \\ \frac{d}{dh}(S^2) &> 0, \quad h > \left(\frac{6b}{\pi}\right)^{\frac{1}{3}}.\end{aligned}$$

$\left(\text{Note also that } \frac{d^2}{dh^2}(S^2) = \frac{58b^2}{h^4} > 0 \ \forall \ h.\right)$

SOLUTIONS TO FINAL EXAMINATION #3

(15%) Q1. (a) If $f(x) = \dfrac{1}{8x}$, find $f^{(n)}(x)$, where n is a positive integer.

(b) Consider the function $f(x) = \displaystyle\int_3^{\sqrt{x}} 2t \cos t dt$, $x \in [0, \pi]$.

(i) For what values of x does this function have local extrema?

(ii) Are there any points of inflection?

Solution

(a) $f(x) = \dfrac{1}{8x}$, $f'(x) = -\dfrac{1}{8}x^{-2}$, $f'''(x) = \dfrac{2}{8}x^{-3}$, $f^{(\text{iv})}(x) = -\dfrac{6}{8}x^{-4}, \ldots$

Notice: The derivatives alternate in sign, the power decreases by one each time and the coefficient changes according to a factorial.

i.e.

$$f^{(n)}(x) = \frac{(-1)^n}{8} n! x^{-(n+1)}, \quad n = 1, 2, 3, \ldots$$

(b) Write $f(x) = y = \displaystyle\int_3^u 2t \cos t dt$, $u = \sqrt{x}$.

Then

$$f'(x) = \frac{dy}{dx} = \frac{dy}{du}\frac{du}{dx} = \frac{d}{du}\int_3^u 2t \cos t dt \cdot \frac{1}{2\sqrt{x}}$$

$$= 2u \cos u \cdot \frac{1}{2\sqrt{x}} \quad \text{(Fundamental Theorem of Calculus)}$$

i.e.

$$f'(x) = 2\sqrt{x}\frac{\cos\sqrt{x}}{2\sqrt{x}} = \cos\sqrt{x}, \ x \in (0, \pi],$$

and

$$f''(x) = -\frac{1}{2\sqrt{x}} \sin\sqrt{x}, \ x \in (0, \pi].$$

(Note that f is not differentiable at $x = 0$).

Now,

$$f'(x) = \cos\sqrt{x} = 0$$

$$\Leftrightarrow \ \sqrt{x} = \frac{\pi}{2}$$

$$\text{i.e.} \quad x = \frac{\pi^2}{4}$$

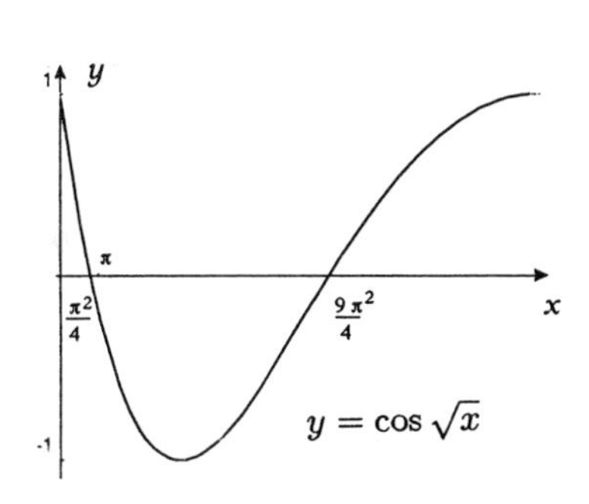

(All other values of x satisfying $f'(x) = 0$ lie outside $(0, \pi]$.)

Also, since $\sin\sqrt{x} > 0 \ \forall \ x \in (0, \pi]$,

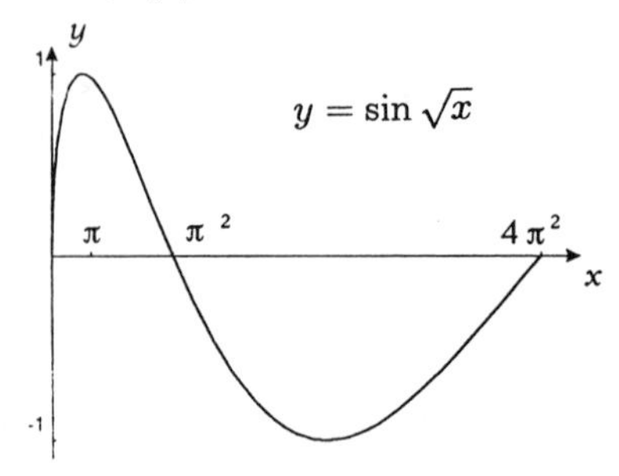

$$f''(x) = -\tfrac{1}{2\sqrt{x}} \sin\sqrt{x} < 0, \ \ x \in (0, \pi]. \text{ i.e.}$$

(i) There is a local maximum at $x = \dfrac{\pi^2}{4}$

and

(ii) The curve is concave down for $x \in (0, \pi]$. There are no points of inflection in this interval.

(30%) Q2. Find: (a) $\displaystyle\int \sin^3 x dx$

(b) $\displaystyle\int \left[\sqrt{\tan\theta} + (\tan\theta)^{\frac{5}{2}}\right] d\theta$

(c) $\displaystyle\int_0^1 \frac{x^{\frac{1}{2}}}{\sqrt{1 + x^{\frac{1}{2}}}} dx$

(d) $\displaystyle\int x^7 \sqrt{1 + x^2} dx$

(e) $\displaystyle\int \frac{\cos^3 x \sin x}{[1 + \cos^4 x]^{\frac{3}{2}}} dx$

(f) $\displaystyle\int_0^\pi \sqrt{\cos^2 x} dx$

(g) $\displaystyle\lim_{x\to\infty} \frac{2 - \sqrt{x}}{2 + \sqrt{x}}$

(h) $\displaystyle\lim_{\theta\to\infty} 3\theta \sin\left(\tfrac{1}{\theta}\right)$

Solution

(a)

$$I = \int \sin^3 x dx = \int \sin x \sin^2 x dx.$$

Let $u = \cos x, \ du = -\sin x dx.$

$$I = -\int (1 - u^2) du \quad (\text{since} \quad \sin^2 x = 1 - \cos^2 x = 1 - u^2)$$

$$= -u - \frac{u^3}{3} + c = -\left(\cos x - \frac{\cos^3 x}{3}\right) + c$$

(b)

$$I = \int \sqrt{\tan\theta}(1+\tan^2\theta)d\theta$$
$$= \int \sqrt{\tan\theta}\sec^2\theta d\theta$$

Let $u = \tan\theta$. Then $du = \sec^2\theta d\theta$ and

$$I = \int \sqrt{u}du = \frac{2}{3}u^{\frac{3}{2}} + c$$
$$= \frac{2}{3}(\tan\theta)^{\frac{3}{2}} + c$$

(c)

$$I = \int_0^1 \frac{x^{\frac{1}{2}}}{\sqrt{1+x^{\frac{1}{2}}}}dx$$

Let $u = 1 + x^{\frac{1}{2}}$; $du = \frac{1}{2x^{\frac{1}{2}}}dx$

$$x = 0, \quad u = 1$$
$$x = 1, \quad u = 2$$

Now,

$$\frac{x^{\frac{1}{2}}}{\sqrt{1+x^{\frac{1}{2}}}}dx = \frac{(u-1)}{\sqrt{u}} \cdot 2(u-1)du = \frac{2(u-1)^2}{\sqrt{u}}du$$

Therefore

$$I = 2\int_1^2 \frac{(u^2-2u+1)}{\sqrt{u}}du = 2\int_1^2 \left(u^{\frac{3}{2}} - 2u^{\frac{1}{2}} + u^{-\frac{1}{2}}\right)du$$
$$= 2\left[\frac{2}{5}u^{\frac{5}{2}} - \frac{4}{3}u^{\frac{3}{2}} + 2u^{\frac{1}{2}}\right]_1^2$$
$$= 2\left[\frac{2}{5}2^{\frac{5}{2}} - \frac{4}{3}2^{\frac{3}{2}} + 2\sqrt{2} - \frac{2}{5} + \frac{4}{3} - 2\right]$$
$$= 2\left[\sqrt{2}\left(\frac{8}{5} - \frac{8}{3} + 2\right) - \frac{16}{15}\right]$$
$$= \frac{4}{15}\left(7\sqrt{2} - 8\right)$$

(d)

$$I = \int x^7 \sqrt{1+x^2}dx.$$

Let $u = 1 + x^2$; $du = 2xdx$

$$\begin{aligned}
I &= \frac{1}{2}\int (u-1)^3\sqrt{u}du \qquad (\text{since} \quad x^6 = (u-1)^3) \\
&= \frac{1}{2}\int \sqrt{u}(u^3 - 3u^2 + 3u - 1)du \\
&= \frac{1}{2}\int \left(u^{\frac{7}{2}} - 3u^{\frac{5}{2}} + 3u^{\frac{3}{2}} - u^{\frac{1}{2}}\right) du \\
&= \frac{1}{2}\left[\frac{2}{9}u^{\frac{9}{2}} - \frac{6}{7}u^{\frac{7}{2}} + \frac{6}{5}u^{\frac{5}{2}} - \frac{2}{3}u^{\frac{3}{2}}\right] + c \\
&= \frac{(1+x^2)^{\frac{3}{2}}}{2}\left[\frac{2}{9}(1+x^2)^3 - \frac{6}{7}(1+x^2)^2 + \frac{6}{5}(1+x^2) - \frac{2}{3}\right] + c
\end{aligned}$$

(e) Let $u = 1 + \cos^4 x$. Then $du = -4\cos^3 x \sin x$ and the integral becomes

$$\begin{aligned}
& -\frac{1}{4}\int \frac{du}{u^{\frac{3}{2}}} \\
&= -\frac{1}{4}\left[-2u^{-\frac{1}{2}}\right] + c \\
&= \frac{1}{2}u^{-\frac{1}{2}} + c \\
&= \frac{1}{2\sqrt{1+\cos^4 x}} + c
\end{aligned}$$

(f)

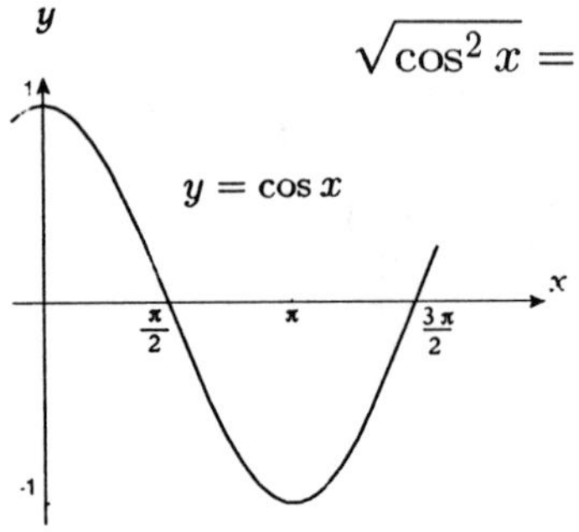

$$\sqrt{\cos^2 x} = |\cos x| = \begin{cases} \cos x, & \cos x \geq 0, \\ -\cos x, & \cos x < 0, \end{cases}$$

$$= \begin{cases} \cos x, & 0 \leq x \leq \frac{\pi}{2}, \\ -\cos x, & \frac{\pi}{2} < x \leq \pi, \end{cases}$$

when $x \in [0, \pi]$.

Hence

$$
\begin{aligned}
\int_0^{\pi} \sqrt{\cos^2 x}dx &= \int_0^{\frac{\pi}{2}} \cos x dx - \int_{\frac{\pi}{2}}^{\pi} \cos x dx \\
&= [\sin x]_0^{\frac{\pi}{2}} - [\sin x]_{\frac{\pi}{2}}^{\pi} \\
&= 1 - (-1) \\
&= 2
\end{aligned}
$$

(g) **Step 2**. Divide top and bottom by $\sqrt{x}$ (highest power present).

$$
\lim_{x\to\infty} \frac{2-\sqrt{x}}{2+\sqrt{x}} = \lim_{x\to\infty} \frac{\frac{2}{\sqrt{x}} - 1}{\frac{2}{\sqrt{x}} + 1}
$$

Step 3

$$
\lim_{x\to\infty} \frac{2-\sqrt{x}}{2+\sqrt{x}} = \frac{\lim\limits_{x\to\infty} \frac{2}{\sqrt{x}} - 1}{\lim\limits_{x\to\infty} \frac{2}{\sqrt{x}} + 1} = -1
$$

(h)

$$
\lim_{\theta\to\infty} 3\theta \sin\left(\frac{1}{\theta}\right)
$$

Stage 1

$$
(\infty \cdot \sin(0) : \quad \text{meaningless!})
$$

Stage 2

Simplify the expression.

Bearing in mind that $\lim\limits_{x\to 0} \dfrac{\sin x}{x} = 1$, write

$$
\lim_{\theta\to\infty} 3\theta \sin\left(\frac{1}{\theta}\right) = 3 \lim_{\theta\to\infty} \frac{\sin \frac{1}{\theta}}{\frac{1}{\theta}}
$$

Let $x = \dfrac{1}{\theta}$. Clearly, as $\theta \to \infty$, $x \to 0$. Hence

$$
3 \lim_{\theta\to\infty} \frac{\sin\left(\frac{1}{\theta}\right)}{\frac{1}{\theta}} = 3 \lim_{x\to 0} \frac{\sin x}{x}
$$

Stage 3

$$3\lim_{x\to 0}\frac{\sin x}{x} = 3\cdot 1 = 3$$

(20%) Q3. Sketch the graph of

$$y = f(x) = x^{\frac{4}{3}} - 4x^{\frac{1}{3}}$$

indicating where the function is increasing, decreasing, concave up, concave down, has local maxima, local minima, inflection points and asymptotes.

Solution

$$\begin{aligned} y = f(x) &= x^{\frac{1}{3}}(x-4) \\ f'(x) &= \frac{4}{3}x^{\frac{1}{3}} - \frac{4}{3}x^{-\frac{2}{3}} \\ &= \frac{4}{3}x^{-\frac{2}{3}}(x-1), \quad x \neq 0. \end{aligned}$$

$$\begin{aligned} f''(x) &= \frac{4}{9}x^{-\frac{2}{3}} + \frac{8}{9}x^{-\frac{5}{3}} \\ &= \frac{4}{9}x^{-\frac{5}{3}}(x+2), \quad x \neq 0. \end{aligned}$$

(a) x-intercepts, $x = 0, 4$

y-intercepts, $y = 0$

(b) No vertical asymptotes.

$$\lim_{x\to\pm\infty} x^{\frac{1}{3}}(x-4) = \pm\infty \quad \text{so no horizontal asymptote}$$

(c)

$$\begin{aligned} f'(x) &> 0 \Leftrightarrow x > 1 && \Leftrightarrow f \text{ is increasing} \\ f'(x) &< 0 \Leftrightarrow x \in (-\infty, 0)\cup(0,1) && \Leftrightarrow f \text{ is decreasing} \end{aligned}$$

(d) $f'(x) = 0 \Leftrightarrow x = 1$

Therefore $f(1) = -3$ is a local minimum.

(e)

$$f''(x) = \frac{4}{9}\frac{(x+2)}{x^{\frac{5}{3}}}$$

$f''(x) > 0 \Leftrightarrow x \in (-\infty, -2) \cup (0, \infty)$ where $f(x)$ is concave up

$f''(x) < 0 \Leftrightarrow x \in (-2, 0)$ where $f(x)$ is concave down

Points of inflection at $x = -2, 0$, i.e. $(-2, 6\sqrt[3]{2})$ and $(0, 0)$.

Sketch.

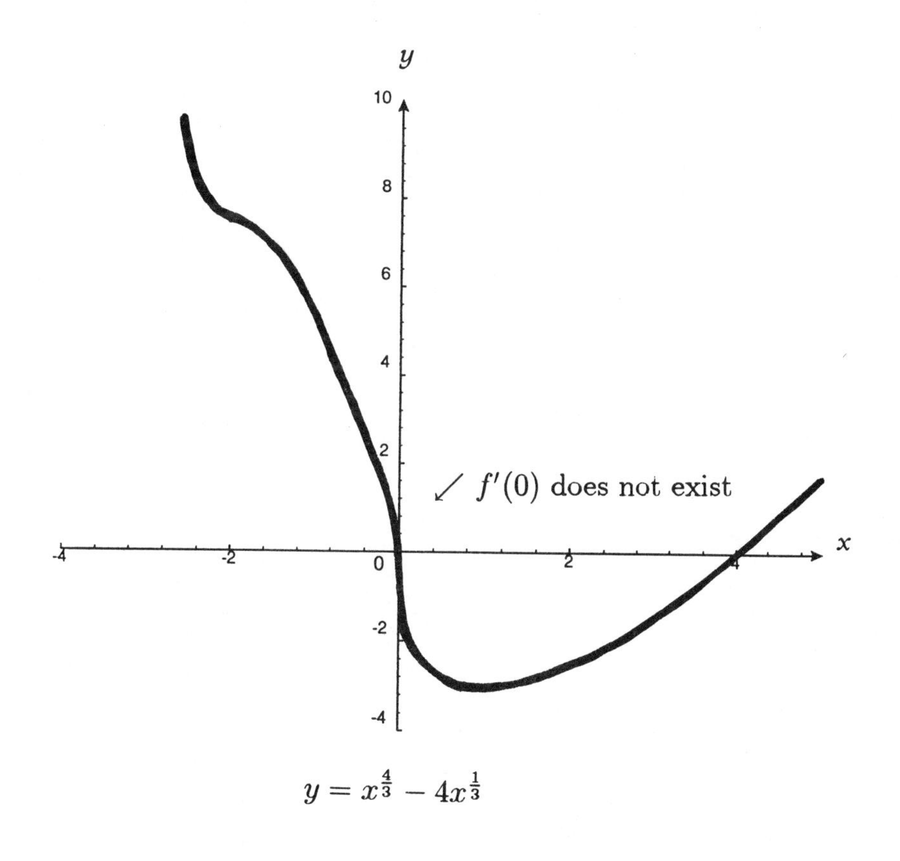

$y = x^{\frac{4}{3}} - 4x^{\frac{1}{3}}$

(8%) Q4. Show that the equation

$$3x^5 + 18x - 1 = 0$$

has *exactly* one real root.

Solution

$$f(x) = 3x^5 + 18x - 1 = 0$$

Since f is continuous and $f(0) = -1$ and $f(1) = 20$, the equation has at least one root in $(0, 1)$ by the Intermediate Value Theorem. Suppose the equation has two roots, $a, b \in \mathbb{R}$ with $a < b$. Then $f(a) = 0 = f(b)$ so by Rolle's Theorem $f'(x) = 15x^4 + 18 = 0$ has a root in (a, b). However, this is impossible since $f'(x) \geq 18$ for all x. Hence $f(x) = 0$ has exactly one real root on the whole real line. In fact, this root lies in $(0, 1)$.

(20%) Q5. Evaluate the following (if they exist) or explain why they don't exist.

(a) $\lim_{x \to 1} \frac{x^{\frac{1}{3}} - 1}{(x - 1)}$

(b) $\lim_{x \to -\infty} \left(\sqrt{2x^2 - 3x + 1} + x\sqrt{2}\right)$

(c) $\lim_{h \to 0} \frac{\sin\left(\frac{\pi}{2} + h\right) - 1}{h}$

(d) $\lim_{x \to -\infty} \frac{3x^3}{\sqrt[3]{x^9 + 2x + 3}}$

Solution

(a) **Stage 1**

$\left(\frac{0}{0}\right)$

Stage 2

Simplify the expression.

$$a^3 - b^3 = (a - b)(a^2 + ab + b^2) \quad a, b \in \mathbb{R} \quad \text{(Section 3.5, Part 1)}$$

Let $a = x^{\frac{1}{3}}$; $b = 1$.

Then

$$\frac{x^{\frac{1}{3}} - 1}{(x - 1)} = \frac{x^{\frac{1}{3}} - 1}{(x - 1)} \cdot \frac{\left(x^{\frac{2}{3}} + x^{\frac{1}{3}} + 1\right)}{\left(x^{\frac{2}{3}} + x^{\frac{1}{3}} + 1\right)}$$

$$= \frac{\left(x^{\frac{1}{3}}\right)^3 - (1)^3}{(x - 1)\left(x^{\frac{2}{3}} + x^{\frac{1}{3}} + 1\right)} = \frac{(x - 1)}{(x - 1)\left(x^{\frac{2}{3}} + x^{\frac{1}{3}} + 1\right)}$$

Stage 3

$$\lim_{x\to1}\frac{x^{\frac{1}{3}}-1}{(x-1)}=\lim_{x\to1}\frac{(x-1)}{(x-1)\left(x^{\frac{2}{3}}+x^{\frac{1}{3}}+1\right)}=\lim_{x\to1}\frac{1}{\left(x^{\frac{2}{3}}+x^{\frac{1}{3}}+1\right)}=\frac{1}{3}$$

(b) **Step 1**

$$\lim_{x\to-\infty}\left(\sqrt{2x^2-3x+1}+\sqrt{2}x\right)$$

$$=\lim_{x\to-\infty}\left(\sqrt{2x^2-3x+1}+\sqrt{2}x\right)\cdot\frac{(\sqrt{2x^2-3x+1}-\sqrt{2}x)}{(\sqrt{2x^2-3x+1}-\sqrt{2}x)}$$

$$=\lim_{x\to-\infty}\frac{2x^2-3x+1-2x^2}{\sqrt{2x^2-3x+1}-\sqrt{2}x}=\lim_{x\to-\infty}\frac{1-3x}{\sqrt{2x^2-3x+1}-\sqrt{2}x}$$

Step 2. Divide top and bottom by x (highest power).

$$\lim_{x\to-\infty}\frac{1-3x}{\sqrt{2x^2-3x+1}-x\sqrt{2}}=\lim_{x\to-\infty}\frac{\frac{1}{x}-3}{\frac{\sqrt{2x^2-3x+1}}{x}-\sqrt{2}}$$

Now, $\sqrt{x^2}=|x|=-x,\ x<0$.

So limit becomes

$$\lim_{x\to-\infty}\frac{\frac{1}{x}-3}{-\sqrt{2-\frac{3}{x}+\frac{1}{x^2}}-\sqrt{2}}\quad\text{(Again}\quad x=-\sqrt{x^2}\quad\text{since}\quad x<0)$$

$$=\frac{-3}{-2\sqrt{2}}=\frac{3}{2\sqrt{2}}$$

(c)

$$\lim_{h\to0}\frac{\sin\left(\frac{\pi}{2}+h\right)-1}{h}=\lim_{h\to0}\frac{\sin\left(\frac{\pi}{2}+h\right)-\sin\left(\frac{\pi}{2}\right)}{h}$$

$$=f'\left(\frac{\pi}{2}\right)\quad\text{where}\quad f(x)=\sin x$$

$$=\cos\left(\frac{\pi}{2}\right)=0$$

(d) **Step 2.** Divide top and bottom by x^3.

$$\lim_{x\to-\infty}\frac{3x^3}{(x^9+2x+3)^{\frac{1}{3}}}=\lim_{x\to-\infty}\frac{3}{\frac{(x^9+2x+3)^{\frac{1}{3}}}{x^3}}$$

Now, $x^3=(x^9)^{\frac{1}{3}}$ so limit becomes

Step 3

$$\lim_{x\to-\infty}\frac{3}{\left(1+\frac{2}{x^8}+\frac{3}{x^9}\right)^{\frac{1}{3}}}=3$$

(7%) Q6. (a) When a sample of gas is compressed at a constant temperature, the pressure $P(t)$ and the volume $V(t)$, at time t, are related by the equation

$$PV=C,\quad C \text{ is constant.}$$

Suppose that at a certain instant $V=10\text{cm}^3$ and $P=8$ Pascals with the latter falling at a rate of 2 Pascals/minute. At what rate is the volume increasing at this instant?

(b) Verify the Mean Value Theorem for the function

$$f(x)=\cos x \quad \text{on} \quad \left[-\frac{\pi}{2},0\right].$$

Solution

(a) **We know:** $PV=C$, C is constant, $P(t),\ V(t)$

$$\frac{dP}{dt}=-2$$

We want: $\dfrac{dV}{dt}$ when $V=10,\ P=8.$

Relate what we *want* to what we *know*:

$$PV=C$$

Therefore

$$P\frac{dV}{dt}+V\frac{dP}{dt}=0$$

i.e.

$$\frac{dV}{dt} = -\frac{V}{P}\frac{dP}{dt} = -\frac{10}{8}(-2)$$
$$= \frac{5}{2}\text{cm}^3/\text{min}$$

(b) $f(x) = \cos x, \quad x \in \left[-\frac{\pi}{2}, 0\right].$

$f(x) = \cos x$ is continuous and differentiable $\forall\, x \in \mathbb{R}$, so it is certainly continuous on $\left[-\frac{\pi}{2}, 0\right]$ and differentiable on $\left(-\frac{\pi}{2}, 0\right)$.

Hence, by the Mean Value Theorem, $\exists$ a number $c \in \left(-\frac{\pi}{2}, 0\right)$ such that

$$f(0) - f\left(-\frac{\pi}{2}\right) = f'(c)\left(0 - \left(-\frac{\pi}{2}\right)\right)$$

i.e.

$$1 - 0 = -\sin c\left(\frac{\pi}{2}\right)$$

i.e.

$$\sin c = -\frac{2}{\pi}$$

which gives

$$c = -\sin^{-1}\left(\frac{2}{\pi}\right) \quad (\sim -0.69)$$
$$\in \left(-\frac{\pi}{2}, 0\right) \quad \text{as required.}$$

SOLUTIONS TO FINAL EXAMINATION #4

(10%) Q1. Determine all line asymptotes of each of the following functions and sketch the graph. (Note: no derivatives are required.)

(i) $f(x) = \dfrac{6x}{x-2}$

(ii) $g(x) = \dfrac{x^2-2}{x+1}$

Solution

(i)

$$f(x) = \frac{6x}{x-2} = 6 + \frac{12}{x-2}$$

$$\lim_{x \to \pm\infty} f(x) = 6$$

Therefore horizontal asymptote at $y = 6$.

Vertical asymptote at $x = 2$.

$$\left.\begin{aligned} \lim_{x \to 2^-} \frac{6x}{x-2} &= \frac{12}{\text{small negative no.}} = -\infty \\ \lim_{x \to 2+} \frac{6x}{x-2} &= \frac{12}{\text{small positive no.}} = +\infty \end{aligned}\right\}$$

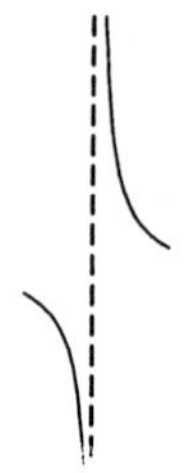

$x = 2$

Intercepts: $x = 0$, $y = 0$.

Sketch:

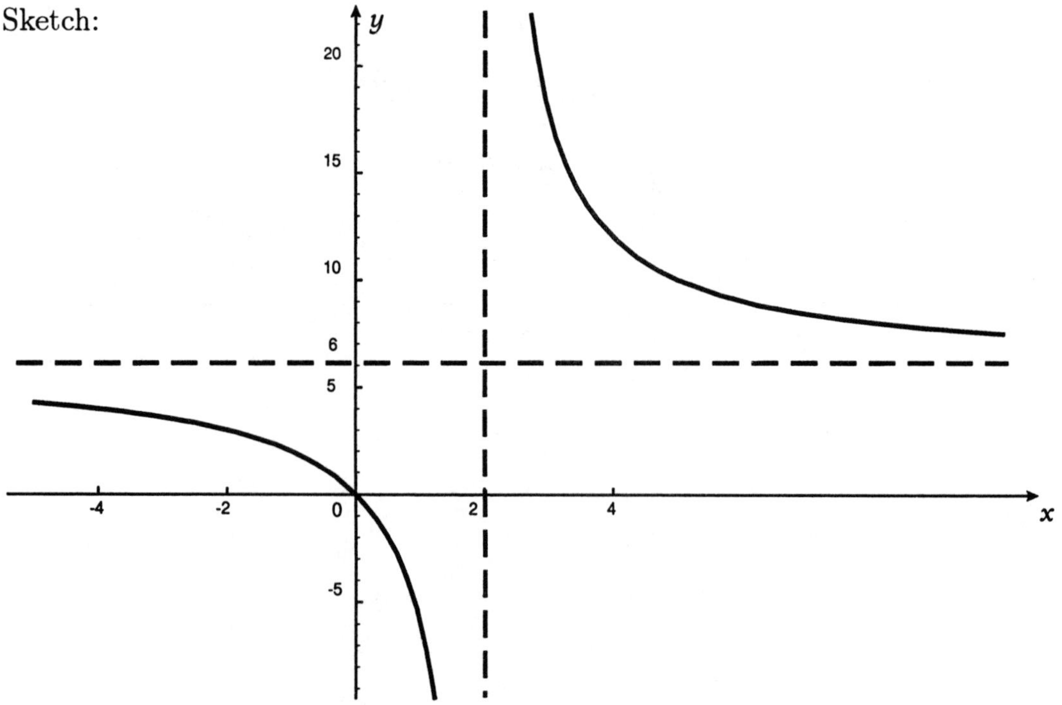

(ii)

$$g(x) = \frac{x^2 - 2}{x + 1} \qquad \text{Divide (see Section 5.2, Part 1).}$$

$$\begin{array}{r|l} & \;\; x - 1 \\ \hline x + 1 & x^2 - 2 \\ & \underline{x^2 + x} \\ & \quad -2 - x \\ & \quad \underline{-1 - x} \\ & \quad -1 \end{array}$$

Therefore

$$g(x) = x - 1 - \frac{1}{x + 1}$$

$$\lim_{x \to \pm\infty} (g(x) - (x - 1)) = 0$$

Therefore $y = x - 1$ is a slant asymptote.

Vertical asymptote at $x = -1$

$$\left.\begin{array}{l} \displaystyle\lim_{x \to -1^-} \frac{x^2 - 2}{x + 1} = \frac{-1}{\text{(small negative no.)}} = +\infty \\ \displaystyle\lim_{x \to -1^+} \frac{x^2 - 2}{x + 1} = \frac{-1}{\text{(small positive no.)}} = -\infty \end{array}\right\}$$

$x = -1$

Intercepts:

$$x = 0, \quad y = -2$$

$$y = 0, \quad x = \pm\sqrt{2}$$

Sketch:

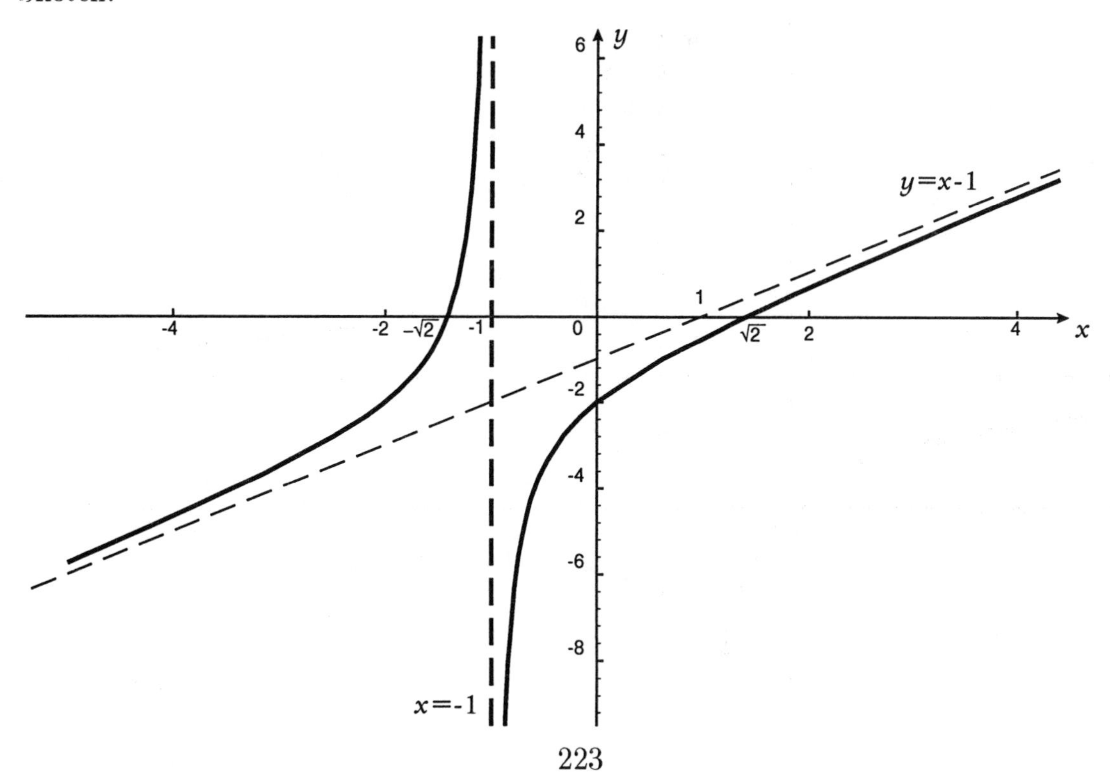

(15%) Q2. Suppose a spherical ball loses air in such a way that its volume decreases at a rate proportional to its surface area. If it takes the ball 4 hours to deflate to half its original volume, how much longer will it take for the ball to deflate completely?

Solution

The volume of a sphere of radius r is $V = \frac{4}{3}\pi r^3$ and the corresponding surface area is $A = 4\pi r^2$. Let t represent time. Clearly $V(t)$, $A(t)$ and $r(t)$.

At $t = 0$,

$$V = V_0$$
$$A = A_0$$

We know: (i) $\dfrac{dV}{dt} = kA = k4\pi r^2, \; k = \text{constant} < 0$.

(ii) At $t = 4$, $V = \dfrac{1}{2}V_0$.

We want: t such that $V = 0$ i.e. such that $r = 0$.

Now,

$$V = \frac{4}{3}\pi r^3 \Rightarrow \frac{dV}{dt} = \frac{dV}{dr}\frac{dr}{dt} = 4\pi r^2 \frac{dr}{dt}$$
$$= A\frac{dr}{dt}$$

But, we are given $\dfrac{dV}{dt} = kA$. Hence, $\dfrac{dr}{dt} = k$.

Thus, $r = kt + c$, c is an arbitrary constant.

Hence,

$$\text{at} \quad t = 0, \quad r = c$$
$$\text{at} \quad t = 4, \quad r = 4k + c$$

Further,

$$\text{at} \quad t = 4, \quad V = \frac{1}{2}V_0$$

i.e.

$$\frac{4}{3}\pi \underbrace{(4k + c)}_{r \text{ at } t=4}{}^3 = \frac{1}{2} \cdot \frac{4}{3}\pi \underbrace{(c)}_{r \text{ at } t=0}{}^3$$

i.e.

$$(4k + c)^3 = \frac{1}{2}(c)^3$$

i.e.

$$4k + c = \frac{1}{2^{\frac{1}{3}}} c$$

Therefore

$$k = \frac{c}{4}\left(\frac{1}{2^{\frac{1}{3}}} - 1\right)$$

Thus

$$r = \frac{c}{4}\left(\frac{1}{2^{\frac{1}{3}}} - 1\right) t + c$$

Hence,

$$r = 0 \Leftrightarrow \frac{c}{4}\left(\frac{1}{2^{\frac{1}{3}}} - 1\right) t + c = 0$$

i.e.

$$t = \frac{4 \cdot 2^{\frac{1}{3}}}{2^{\frac{1}{3}} - 1} \approx 19.4 \text{ hours}$$

So it takes $\dfrac{4 \cdot 2^{\frac{1}{3}}}{2^{\frac{1}{3}} - 1} - 4$ hours longer, i.e. ~ 15.4 hours longer for the ball to deflate completely.

(15%) Q3. Find the following limits or explain why they don't exist.

(a) $\displaystyle \lim_{x \to 0} \frac{2x}{|x-2| - |x+2|}$

(b) $\displaystyle \lim_{x \to -2} \left(\frac{1}{x+2} - \frac{23}{(x^2-4)}\right)$

(c) $\displaystyle \lim_{x \to a} \frac{f(x) - f(a)}{\sqrt[3]{x} - \sqrt[3]{a}}, \; a > 0$

Solution

(a) **Stage 1**

$\left(\frac{0}{0}\right)$

Stage 2

Simplify the expression

$$|x-2| = \begin{cases} x-2, & x \geq 2, \\ 2-x, & x < 2, \end{cases}$$

$$|x+2| = \begin{cases} x+2, & x \geq -2, \\ -(x+2), & x < -2, \end{cases}$$

Hence,

$$\lim_{x\to 0} \frac{2x}{|x-2|-|x+2|} = \lim_{x\to 0} \frac{2x}{2-x-(x+2)} = \lim_{x\to 0} \frac{2x}{-2x}$$

(since 'near $x = 0$',

$$|x-2| = 2-x$$
$$|x+2| = 2+x)$$

Stage 3

$$\lim_{x\to 0} \frac{2x}{-2x} = -1$$

(b) **Stage 1**

Write

$$\frac{1}{x+2} - \frac{23}{x^2-4} = \frac{1}{(x+2)} - \frac{23}{(x+2)(x-2)}$$
$$= \frac{(x-2)-23}{(x+2)(x-2)} = \frac{x-25}{(x+2)(x-2)}$$

Now, $\lim_{x\to -2} \frac{x-25}{(x+2)(x-2)}$ is of the form $\left(\frac{-27}{0}\right)$.

So the limit doesn't exist, i.e. the denominator gets smaller while the numerator tends to a fixed quantity, i.e. -27. Thus the expression grows indefinitely.

(c) **Stage 1**

$\left(\frac{0}{0}\right)$

Stage 2

Simplify the expression

$$\frac{f(x)-f(a)}{x^{\frac{1}{3}}-a^{\frac{1}{3}}} = \frac{(f(x)-f(a))}{(x^{\frac{1}{3}}-a^{\frac{1}{3}})} \cdot \frac{(x^{\frac{2}{3}}+x^{\frac{1}{3}}a^{\frac{1}{3}}+a^{\frac{2}{3}})}{(x^{\frac{2}{3}}+x^{\frac{1}{3}}a^{\frac{1}{3}}+a^{\frac{2}{3}})}$$

(Since $(c^3-d^3) = (c-d)(c^2+cd+d^2)$ and let $c = x^{\frac{1}{3}}$, $d = a^{\frac{1}{3}}$.)

Hence

$$\lim_{x\to a}\frac{f(x)-f(a)}{x^{\frac{1}{3}}-a^{\frac{1}{3}}}=\lim_{x\to a}\frac{[f(x)-f(a)]\left(x^{\frac{2}{3}}+x^{\frac{1}{3}}a^{\frac{1}{3}}+a^{\frac{2}{3}}\right)}{\left(\left(x^{\frac{1}{3}}\right)^3-\left(a^{\frac{1}{3}}\right)^3\right)}$$

$$=\lim_{x\to a}\frac{f(x)-f(a)}{(x-a)}\cdot\left(x^{\frac{2}{3}}+x^{\frac{1}{3}}a^{\frac{1}{3}}+a^{\frac{2}{3}}\right) \qquad (*)$$

Stage 3

If f *is differentiable at* $x=a$, writing $h=x-a$,

$$f'(a)=\lim_{h\to 0}\frac{f(a+h)-f(a)}{h}=\lim_{x\to a}\frac{f(x)-f(a)}{x-a}\quad \text{is finite.}$$

Hence, from (*),

$$\lim_{x\to a}\frac{f(x)-f(a)}{x^{\frac{1}{3}}-a^{\frac{1}{3}}}=f'(a)\cdot 3a^{\frac{2}{3}}$$

If f *is not differentiable at* $x=a$, then $f'(a)$ does not exist and from (*) $\lim_{x\to a}\frac{f(x)-f(a)}{x^{\frac{1}{3}}-a^{\frac{1}{3}}}$ doesn't exist.

(30%) Q4. Find the following.

(a) $\int \frac{3x^6-2x^{\frac{1}{2}}}{x^4}dx$

(b) $\int \frac{x+\sqrt{x^2+2}}{\sqrt{x^2+2}}dx$

(c) $\int \cos x\sec^2(\sin x)dx$

(d) $\int_{-\frac{\pi}{3}}^{\frac{\pi}{3}} \sin^7\theta d\theta$

(e) $f''(x)$ if $f(x)=\int_0^{\sqrt{x}}\left[\int_{\frac{1}{2}}^{\cos t}(1+u^2)^{\frac{3}{2}}du\right]dt$

(f) $\int_0^2 |x^2-3x+2|dx$

Solution

(a)

$$I=\int\left(3x^2-2x^{-\frac{7}{2}}\right)dx$$

$$=x^3+\frac{4}{5}x^{-\frac{5}{2}}+c$$

(b)

$$\begin{aligned} I &= \int \left(1 + \frac{x}{\sqrt{x^2+2}}\right) dx = x + \int \frac{x}{\sqrt{x^2+2}} dx \\ &= x + \frac{1}{2}\int u^{-\frac{1}{2}} du \quad (\text{letting} \quad u = x^2+2) \\ &= x + \sqrt{x^2+2} + c \end{aligned}$$

(c) Let $u = \sin x,\ du = \cos x dx$

$$\begin{aligned} I = \int \sec^2 u du &= \tan u + c \\ &= \tan(\sin x) + c \end{aligned}$$

(d)

$$I = \int_{-\frac{\pi}{3}}^{\frac{\pi}{3}} \sin^7 \theta d\theta = 0$$

(I is of the form $\displaystyle\int_{-a}^{a} f(\theta)d\theta$ where f is odd, i.e. $f(-\theta) = -f(\theta)$).

(e)

$$f'(x) = \frac{d}{dx}\int_0^{\sqrt{x}} \left[\int_{\frac{1}{2}}^{\cos t} (1+u^2)^{\frac{3}{2}} du\right] dt = \frac{d}{dx}\int_0^{\sqrt{x}} g(t)dt$$

where $g(t) = \displaystyle\int_{\frac{1}{2}}^{\cos t} (1+u^2)^{\frac{3}{2}} du.$

Hence

$$\begin{aligned} f'(x) &= \frac{d}{dw}\int_0^{w} g(t)dt \ \frac{dw}{dx} \qquad (w = \sqrt{x}) \\ &= g(w) \cdot \frac{1}{2\sqrt{x}} \qquad (\text{Fundamental Theorem of Calculus}) \\ &= \int_{\frac{1}{2}}^{\cos\sqrt{x}} (1+u^2)^{\frac{3}{2}} du \cdot \frac{1}{2\sqrt{x}} \end{aligned}$$

Using the product rule

$$f''(x) = \frac{1}{2\sqrt{x}} \cdot \frac{d}{dx} \int_{\frac{1}{2}}^{\cos\sqrt{x}} (1+u^2)^{\frac{3}{2}}\, du - \frac{1}{4x^{\frac{3}{2}}} \int_{\frac{1}{2}}^{\cos\sqrt{x}} (1+u^2)^{\frac{3}{2}}\, du.$$

Now, by the Fundamental Theorem and chain rule,

$$\begin{aligned} \frac{d}{dx} \int_{\frac{1}{2}}^{\cos\sqrt{x}} (1+u^2)^{\frac{3}{2}}\, du &= \frac{d}{ds} \int_{\frac{1}{2}}^{s} (1+u^2)^{\frac{3}{2}}\, du \cdot \frac{ds}{dx} \qquad (s = \cos\sqrt{x}) \\ &= (1+s^2)^{\frac{3}{2}} \left(-\frac{1}{2\sqrt{x}} \sin\sqrt{x}\right) \\ &= (1+\cos^2\sqrt{x})^{\frac{3}{2}} \left(-\frac{1}{2\sqrt{x}} \sin\sqrt{x}\right) \end{aligned}$$

Hence,

$$f'(x) = -\frac{1}{4x} \sin\sqrt{x}\,(1+\cos^2\sqrt{x})^{\frac{3}{2}} - \frac{1}{4x^{\frac{3}{2}}} \int_{\frac{1}{2}}^{\cos\sqrt{x}} (1+u^2)^{\frac{3}{2}}\, du$$

(f) Using the test-point method (Section 3.8, Part 1),

$$|x^2-3x+2| = |(x-1)(x-2)| = \begin{cases} x^2-3x+2, & x \le 1,\ x \ge 2, \\ -(x^2-3x+2), & 1 < x < 2. \end{cases}$$

$$\begin{aligned} I &= \int_0^1 (x^2-3x+2)\, dx - \int_1^2 (x^2-3x+2)\, dx \\ &= \left[\frac{x^3}{3} - \frac{3x^2}{2} + 2x\right]_0^1 - \left[\frac{x^3}{3} - \frac{3x^2}{2} + 2x\right]_1^2 \end{aligned}$$

Therefore

$$\begin{aligned} I &= \left(\frac{1}{3} - \frac{3}{2} + 2\right) - \left[\left(\frac{8}{3} - 6 + 4\right) - \left(\frac{1}{3} - \frac{3}{2} + 2\right)\right] \\ &= \frac{2}{3} - 3 + 4 - \frac{8}{3} + 6 - 4 \\ &= 1 \end{aligned}$$

(10%) Q5. Suppose a piece of wire of length 2m is cut into two pieces one of which is bent into an equilateral triangle and the other into a circle. How should the wire be cut so that the sum of the areas is a

(i) maximum?

(ii) minimum?

Solution

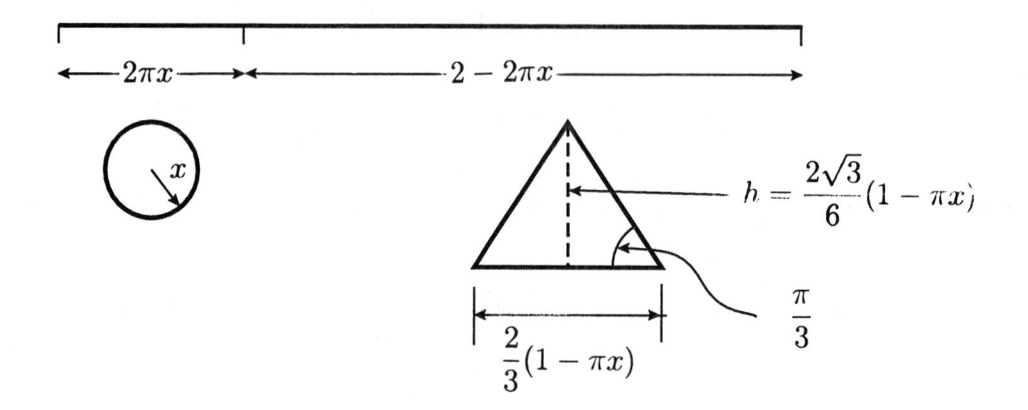

$$\text{Total Area:} \quad A(x) = \underbrace{\pi x^2}_{\text{circle}} + \underbrace{\frac{1}{2} 2\left(\frac{1-\pi x}{3}\right) \cdot \frac{\sqrt{3}\cdot 2(1-\pi x)}{6}}_{\text{triangle } \left(\frac{1}{2} \text{ base * height}\right)}$$

$$= \pi x^2 + \frac{\sqrt{3}}{9}(1-\pi x)^2 \qquad 0 \le x \le \frac{1}{\pi}$$

$$A'(x) = 2\pi x - 2\frac{\pi\sqrt{3}}{9}(1-\pi x)$$

$$= x\left(2\pi + 2\pi^2\frac{\sqrt{3}}{9}\right) - \frac{2\pi\sqrt{3}}{9}$$

$$= x2\pi\left(1 + \pi\frac{\sqrt{3}}{9}\right) - \frac{2\pi\sqrt{3}}{9}$$

i.e.

$$A'(x) = 0 \Leftrightarrow x = \frac{2\pi\sqrt{3}}{9\ 2\pi\left(1+\frac{\pi\sqrt{3}}{9}\right)} = \frac{\sqrt{3}}{9+\pi\sqrt{3}}$$

$$A\left(\frac{\sqrt{3}}{9+\pi\sqrt{3}}\right) = \frac{\pi 3}{(9+\pi\sqrt{3})^2} + \frac{\sqrt{3}}{9}\left(1 - \frac{\pi\sqrt{3}}{9+\pi\sqrt{3}}\right)^2 \sim 0.1198$$

Now, at the endpoints,

$$A(0) = \frac{\sqrt{3}}{9}, \quad A\left(\frac{1}{\pi}\right) = \frac{1}{\pi}$$

Hence,

(i) The maximum occurs when $x = \frac{1}{\pi}$m, i.e. whole wire is bent into a circle radius $\frac{1}{\pi}$m.

(ii) The minimum occurs when $x = \frac{\sqrt{3}}{9+\pi\sqrt{3}} \sim 0.119$m, i.e. wire is cut into 2 pieces – one of length $2\pi\frac{\sqrt{3}}{9+\pi\sqrt{3}} \sim 0.748$m which forms the circle and the other of length ~ 1.252m which forms the equilateral triangle of side ~ 0.417m.

(10%) Q6. (a) Find y' if $x^2y^3 + \cos\sqrt{xy} = \sqrt{x\sqrt{x}}$. (Do not simplify your answer.)

(b) Does continuity of a function at a point imply differentiability at that point? Illustrate with an example.

Solution

(a) Implicit differentiation

$$2xy^3 + 3y^2y'x^2 - \sin\sqrt{xy}\left(\frac{1}{2\sqrt{xy}}(xy' + y)\right) = \frac{3}{4}x^{-\frac{1}{4}}$$

$$\left(\text{Since } \sqrt{x\sqrt{x}} = \sqrt{x^{\frac{3}{2}}} = x^{\frac{3}{4}}\right).$$

Hence,

$$y'\left[3x^2y^2 - \frac{x\sin\sqrt{xy}}{2\sqrt{xy}}\right] = \frac{3}{4}x^{-\frac{1}{4}} - 2xy^3 + \frac{y\sin\sqrt{xy}}{2\sqrt{xy}}$$

and

$$y' = \frac{\frac{3}{4}x^{-\frac{1}{4}} - 2xy^3 + \frac{y\sin\sqrt{xy}}{2\sqrt{xy}}}{3x^2y^2 - \frac{x\sin\sqrt{xy}}{2\sqrt{xy}}}$$

(b) No! Consider

$$y = f(x) = |x| = \begin{cases} x, & x \geq 0, \\ -x, & x < 0 \end{cases}$$

This function is clearly continuous at $x = 0$.

i.e.

$$\lim_{x \to 0^-} f(x) = \lim_{x \to 0^-} (-x) = 0$$
$$\lim_{x \to 0+} f(x) = \lim_{x \to 0+} (x) = 0$$
$$\lim_{x \to 0} f(x) = 0 = f(0).$$

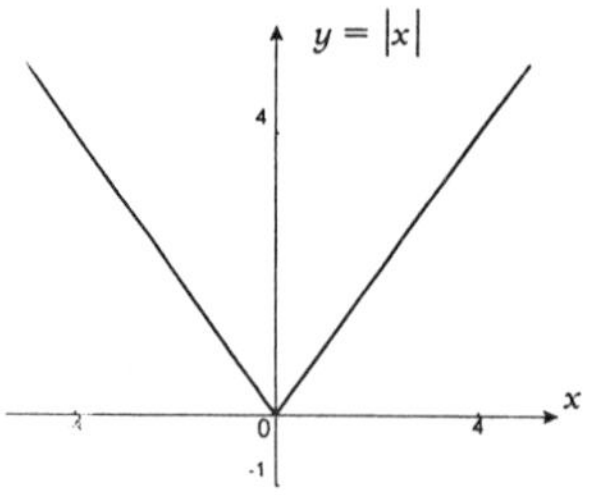

However,

$$f'(0) = \lim_{h \to 0} \frac{f(0+h) - f(0)}{h} = \lim_{h \to 0} \frac{|h|}{h}$$

Thus,

$$\lim_{h \to 0+} \frac{f(0+h) - f(0)}{h} = 1$$
$$\lim_{h \to 0^-} \frac{f(0+h) - f(0)}{h} = -1$$

Thus $f'(0)$ does not exist (since the one-sided limits are not equal) and so $f(x)$ is not differentiable at $x = 0$ although continuous there.

(The graph of $y = |x|$ illustrates the problem with differentiability at $x = 0$ i.e. the "corner" at $x = 0$ means that there can be no unique tangent to the curve at this point.)

(10%) Q7. Find the total area bounded by the curves $y = x$ and $y = x^{11}$.

Solution

Curves intersect when $x^{11} - x = 0$, i.e. $x(x^{10} - 1) = 0$ i.e. $x = 0, \pm 1$

$$
\begin{aligned}
A = \int_{-1}^{1} |x^{11} - x| dx &= \int_{-1}^{0} (x^{11} - x) dx + \int_{0}^{1} (x - x^{11}) dx \\
&= 2 \int_{0}^{1} (x - x^{11}) dx \\
&= 2 \left[\frac{x^2}{2} - \frac{x^{12}}{12} \right]_0^1 \\
&= 2 \left(\frac{1}{2} - \frac{1}{12} \right) = \frac{5}{6}
\end{aligned}
$$

Sketch

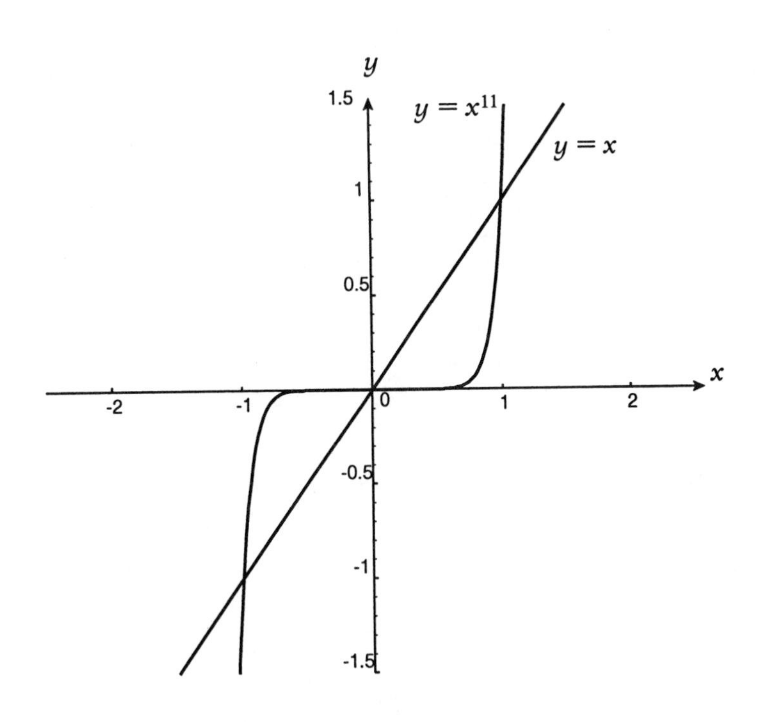

SOLUTIONS TO FINAL EXAMINATION #5

(25%) Q1. Find

(a) $\lim_{x\to 0+} \dfrac{\sqrt{1-\cos 2x}}{x\sec x}$ **OR*** $\lim_{x\to 0} \dfrac{3^x - 1}{x}$

(b) $\lim_{x\to\infty} \dfrac{2-\sqrt{x}}{2+\sqrt{x}}$ **OR*** $\lim_{x\to\infty} e^{-x}\ln x$

(c) $\lim_{x\to\infty} \sin\left(\dfrac{1}{x}\right)$ **OR*** $\lim_{x\to -\infty} \dfrac{e^{3x}-e^{-3x}}{e^{3x}+e^{-3x}}$

(d) $\lim_{x\to 3^+} \left|\dfrac{x-3}{x+1}\right|$

(e) $f^{(n)}(x)$, $n \geq 1$, if $f(x) = \dfrac{x+1}{x}$

Solution

(a) **Stage 1**

$\left(\dfrac{0}{0}\right)$

Stage 2

$$\sqrt{1-\cos 2x} = \sqrt{2}\sin x \qquad (\text{since } \cos 2x = 1 - 2\sin^2 x)$$

Hence

$$\lim_{x\to 0+} \frac{\sqrt{1-\cos 2x}}{x\sec x} = \frac{\sqrt{2}\sin x\cos x}{x}$$

Stage 3

$$\lim_{x\to 0+} \frac{\sqrt{2}\sin x\cos x}{x} = \sqrt{2}\lim_{x\to 0+}\frac{\sin x}{x} \cdot \lim_{x\to 0+}\cos x$$
$$= \sqrt{2}\cdot 1\cdot 1 = \sqrt{2}$$

OR*

$$\lim_{x\to 0}\frac{3^x-1}{x} \qquad \left(\text{indeterminate of form } \left(\frac{0}{0}\right)\right)$$
$$\underset{\text{l'Hôpital}}{=} \lim_{x\to 0}\frac{3^x\ln 3}{1}$$
$$= \ln 3$$

(b)

$$\lim_{x\to\infty}\frac{2-\sqrt{x}}{2+\sqrt{x}}=\lim_{x\to\infty}\frac{\frac{2}{\sqrt{x}}-1}{\frac{2}{\sqrt{x}}+1}=-1\quad\text{(Dividing top and bottom by }\sqrt{x}\text{)}$$

OR*

$$\lim_{x\to\infty}\frac{\ln x}{e^x}\qquad\left(\text{indeterminate of form }\left(\frac{0}{0}\right)\right)$$

$$\underset{\text{l'Hôpital}}{=}\lim_{x\to\infty}\frac{\frac{1}{x}}{e^x}$$

$$=\lim_{x\to\infty}\frac{1}{xe^x}$$

$$=0$$

(c) Let $u=\dfrac{1}{x}$. Then as $x\to\infty$, $u\to 0$. Hence,

$$\lim_{x\to\infty}\sin\left(\frac{1}{x}\right)=\lim_{u\to 0}\sin u=0$$

OR*

$$\lim_{x\to-\infty}\frac{e^{3x}-e^{-3x}}{e^{3x}+e^{-3x}}$$

$$=\lim_{x\to-\infty}\frac{e^{6x}-1}{e^{6x}+1}\qquad\left(\text{Dividing top and bottom by }e^{-3x}\right)$$

$$=-1$$

(d) **Stage 1**

$\left(\dfrac{0}{4}\right)$

Hence,

$$\lim_{x\to 3^+}\left|\frac{x-3}{x+1}\right|=0$$

(i.e. $f(x) = \left|\dfrac{x-3}{x+1}\right|$ is continuous at $x = 3$)

(e)

$$f(x) = 1 + \frac{1}{x}$$

$$f'(x) = -\frac{1}{x^2};\; f''(x) = \frac{2}{x^3};\; f'''(x) = -\frac{6}{x^4}$$

$$f^{iv}(x) = \frac{24}{x^5} \cdots$$

$$f^{(n)}(x) = \frac{(-1)^n n!}{x^{n+1}},\quad n = 1, 2, 3 \ldots$$

(20%) Q2. Let $f(x) = \dfrac{1}{x-2} - x$ **OR*** $f(x) = xe^{x^2}$.

(a) What is the domain of $f(x)$?

(b) Write down the equations of all line asymptotes associated with the curve $y = f(x)$.

(c) Where is the graph of f increasing and where decreasing?

(d) Where is the graph of f concave up and where concave down?

(e) Locate any local extrema and any inflection points.

(f) Sketch the curve represented by $y = f(x)$.

Solution

$$y = f(x) = \frac{1}{x-2} - x$$

(a) Remove values of x which lead to $f(x)$ being undefined.

i.e. $x = 2$.

Therefore domain of f is $\{x \in \mathbb{R} : x \neq 2\}$.

(b)

$$\lim_{x \to \pm\infty} (y + x) = 0$$

Hence $y = -x$ is a slant asymptote.

$x = 2$ is a vertical asymptote.

$$\left.\begin{aligned}\lim_{x \to 2^-} f(x) &= \frac{1}{\text{(small negative no.)}} - 2 = -\infty \\ \lim_{x \to 2^+} f(x) &= \frac{1}{\text{(small positive no.)}} - 2 = +\infty\end{aligned}\right\}$$

(c)

$$f'(x) = -1 - \frac{1}{(x-2)^2} < 0 \; \forall \; x \neq 2$$

Hence f is decreasing on $(-\infty, 2) \cup (2, \infty)$.

(d)

$$f''(x) = \frac{2}{(x-2)^3} > 0 \Leftrightarrow x > 2$$

Hence, f is concave up on $(2, \infty)$ and concave down on $(-\infty, 2)$.

(e)

$$f'(x) \neq 0 \quad \text{therefore no local extrema}$$

From (d), no inflection points ($x = 2$ is a vertical asymptote)

(f) Sketch

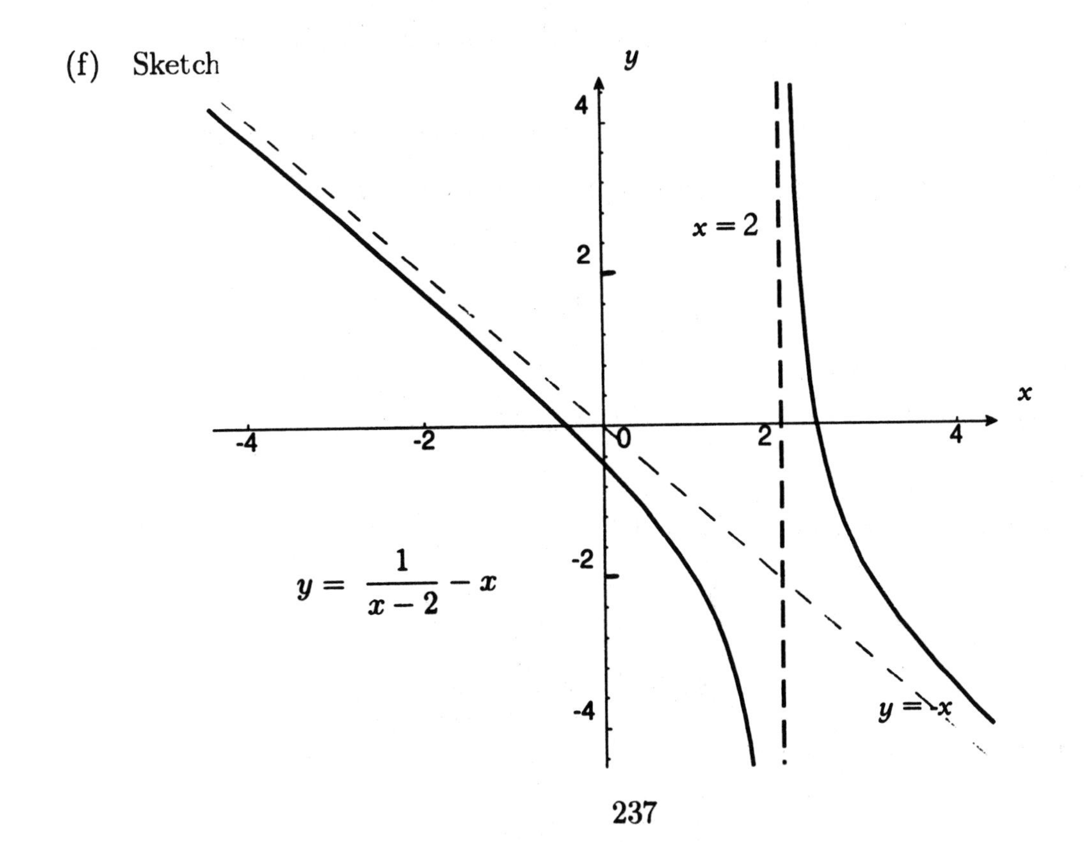

OR*

$$f(x) = xe^{x^2}$$

(a) $f(x)$ exists $\forall x$ so domain is $\mathbb{R}$.

(b) $\lim_{x \to \infty} xe^{x^2} = \infty, \quad \lim_{x \to -\infty} xe^{x^2} = \infty$, no asymptotes

(c)

$$\begin{aligned} f'(x) &= e^{x^2} + (2x)xe^{x^2} \\ &= e^{x^2}(1 + 2x^2) > 0. \end{aligned}$$

So f is increasing on $\mathbb{R}$.

(d)

$$\begin{aligned} f''(x) &= e^{x^2}(2x)(1 + 2x^2) + e^{x^2}(4x) \\ &= e^{x^2}(2x)(3 + 2x^2) \\ &> 0 \Leftrightarrow x > 0 \end{aligned}$$

So f is concave up on $(0, \infty)$ and concave down on $(-\infty, 0)$.

(e) No extrema since $f'(x) > 0 \ \forall x$.

Inflection point at $(0, 0)$.

(f)

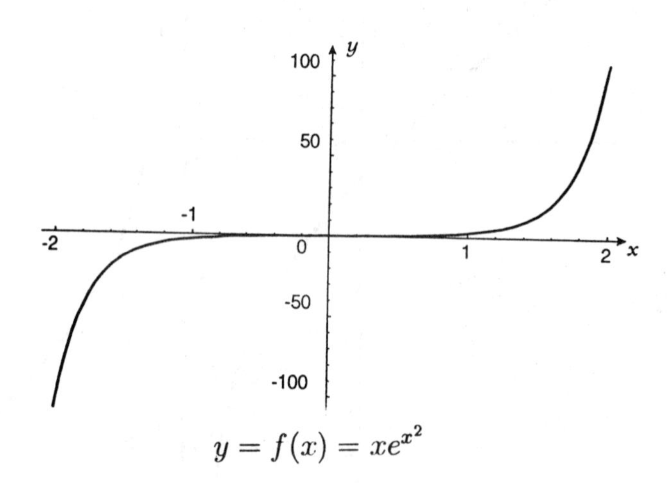

$y = f(x) = xe^{x^2}$

(25%) Q3. Find the following:

(a) $\displaystyle\int \frac{(\sqrt{\sin x} + \sin^2 x)}{\sqrt{1-\cos 2x}} \cos x dx,\ \sin x > 0$

OR* $\displaystyle\int \frac{\sin(\ln x^2)}{x} dx$

(b) $\displaystyle\int \sin^3 t \cos^2 t dt$ **OR*** $\displaystyle\int \frac{x}{x+1} dx$

(c) $\displaystyle\frac{d}{du}\int_u^{3u} \cos\sqrt{1+x^2}dx$

OR* $\displaystyle\frac{d}{dt}\int_1^{\ln t} \sin\sqrt{u^2+e^u}du$

(d) $\displaystyle\int_0^1 \frac{3y^4}{(6+y^5)^8} dy$ **OR*** $\displaystyle\int e^x \sin^2(1+e^x)dx$

(e) $\displaystyle\int_{-1}^2 \sqrt{x^2}dx$ **OR*** $\displaystyle\int \frac{2+e^{-x}}{e^x} dx$

Solution

(a)

$$I = \int \frac{\left(\sqrt{\sin x} + \sin^2 x\right)}{\sqrt{2}\sin x} \cos x dx, \quad \sin x > 0$$

(since $\cos 2x = 1 - 2\sin^2 x$)

Therefore,

$$I = \frac{1}{\sqrt{2}} \int \left(\frac{1}{\sqrt{\sin x}} + \sin x\right) \cos x dx.$$

Let $u = \sin x,\ du = \cos x dx$.

$$\begin{aligned} I &= \frac{1}{\sqrt{2}} \int \left(u^{-\frac{1}{2}} + u\right) du \\ &= \frac{1}{\sqrt{2}}\left(2u^{\frac{1}{2}} + \frac{u^2}{2}\right) + c \\ &= \frac{1}{\sqrt{2}}\left(2\sqrt{\sin x} + \frac{\sin^2 x}{2}\right) + c \end{aligned}$$

OR*

Let $u = \ln x^2 = 2\ln x;\ du = \dfrac{2}{x}dx$

$$\begin{aligned} I = \frac{1}{2}\int \sin u du &= -\frac{1}{2}\cos u + c \\ &= -\frac{1}{2}\cos(\ln x^2) + c \end{aligned}$$

(b) Let $u = \cos t,\ du = -\sin t dt$

$$I = -\int (1 - u^2)u^2 du \quad \text{since} \quad \sin^2 t + \cos^2 t = 1$$

Therefore,

$$\begin{aligned} I &= -\left[\frac{u^3}{3} - \frac{u^5}{5}\right] + c \\ &= -\frac{\cos^3 t}{3} + \frac{\cos^5 t}{5} + c \end{aligned}$$

OR*

$$\begin{aligned} I &= \int \left(1 - \frac{1}{x+1}\right) dx = x - \int \frac{1}{u} du \quad (u = x+1) \\ &= x - \ln|u| + c = x - \ln|x+1| + c \end{aligned}$$

(c)

$$\begin{aligned} \frac{d}{du}\int_u^{3u} \cos\sqrt{1+x^2}du &= \frac{d}{du}\left(\int_u^a + \int_a^{3u}\right) \cos\sqrt{1+x^2}dx \quad (a \text{ is a constant}) \\ &= -\frac{d}{du}\int_a^u \cos\sqrt{1+x^2}dx \\ &\quad + \frac{d}{dw}\int_a^w \cos\sqrt{1+x^2}dx \cdot \frac{dw}{du} \quad (w = 3u) \end{aligned}$$

(Now use Fundamental Theorem.)

$$\begin{aligned} &= -\cos\sqrt{1+u^2} + \cos\sqrt{1+w^2} \cdot 3 \\ &= -\cos\sqrt{1+u^2} + 3\cos\sqrt{1+9u^2} \end{aligned}$$

OR*

$$\begin{aligned} &\frac{d}{dt}\int_1^{\ln t} \sin\sqrt{u^2 + e^u}du \\ &= \frac{d}{d\theta}\int_1^{\theta} \sin\sqrt{u^2+e^u}du \cdot \frac{d\theta}{dt}, \quad \theta = \ln t \\ &= \sin\sqrt{\theta^2 + e^\theta} \cdot \frac{1}{t} \quad \text{(Fundamental Theorem of Calculus)} \\ &= \frac{\sin\sqrt{(\ln t)^2 + t}}{t} \end{aligned}$$

(d) Let $u = y^5 + 6$; $du = 5y^4 dy$,

$$y = 0, \quad u = 6$$
$$y = 1, \quad u = 7$$

$$I = \frac{3}{5}\int_6^7 \frac{du}{u^8} = -\frac{3}{5}\cdot\frac{1}{7}\left[\frac{1}{u^7}\right]_6^7 = -\frac{3}{35}\left(\frac{1}{7^7} - \frac{1}{6^7}\right)$$

OR*

Let $u = 1 + e^x$, $du = e^x dx$

$$I = \int \sin^2 u du = \frac{1}{2}\int (1 - \cos 2u) du \quad (\text{since } \cos 2u = 1 - 2\sin 2u)$$

Therefore

$$I = \frac{1}{2}\left(u - \frac{1}{2}\sin 2u\right) + c$$
$$= \frac{1}{2}\left[1 + e^x - \frac{1}{2}\sin 2(1 + e^x)\right] + c$$

(e)

$$\sqrt{x^2} = |x| = \begin{cases} x, & x \geq 0 \\ -x, & x < 0 \end{cases}$$

Therefore

$$I = \int_{-1}^0 -x dx + \int_0^2 x dx$$
$$= -\left[\frac{x^2}{2}\right]_{-1}^0 + \left[\frac{x^2}{2}\right]_0^2$$
$$= -\left(-\frac{1}{2}\right) + 2 = \frac{5}{2}$$

OR*

Let $u = 2 + e^{-x}$, $du = -e^{-x} dx$

Therefore

$$I = -\int u du = -\frac{u^2}{2} + c$$
$$= -(2 + e^{-x})^2 + c$$

(10%) Q4. A ladder of height h metres rests against a vertical wall. The bottom of the ladder is then pulled along the ground away from the wall at a constant rate of b metres/second. Derive an expression for the speed of the top of the ladder as it slides down the wall in terms of $\theta(t)$ and $\frac{d\theta}{dt}$ where $\theta(t)$ is the angle the ladder makes with the perpendicular to the ground at time t.

Solution

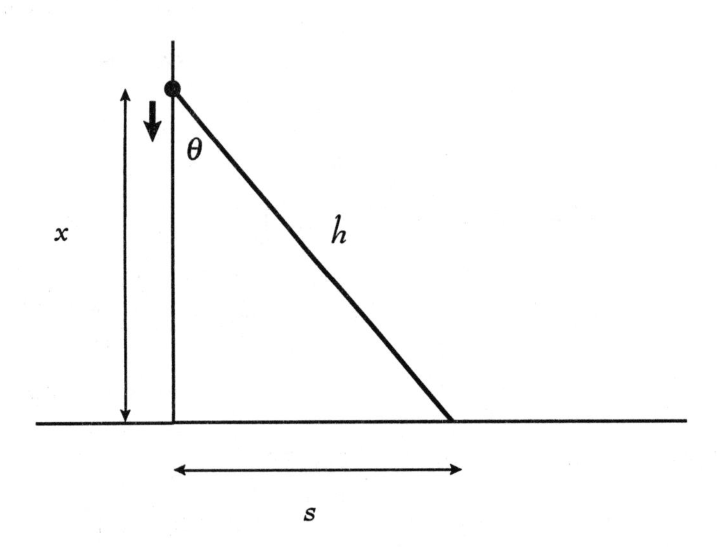

We know: $\frac{ds}{dt} = b\ m/s$

We want: $\frac{dx}{dt}$

Relate what we *want* to what we *know*.

Now,

$$\tan\theta = \frac{s}{x}$$

Therefore,

$$\begin{aligned} x &= \frac{s}{\tan\theta} = s\cot\theta \\ \frac{dx}{dt} &= \cot\theta\frac{ds}{dt} + s(-\csc^2\theta)\frac{d\theta}{dt} \\ &= b\cot\theta - h\sin\theta\csc^2\theta\frac{d\theta}{dt} \quad \left(\text{since } \sin\theta = \frac{s}{h}\right) \\ &= b\cot\theta - h\csc\theta\frac{d\theta}{dt} \end{aligned}$$

(10%) Q5. (a) Consider the ellipse $x^2 + 3y^2 = 1$. Find the equations of the tangent lines with unit slope.

(b) Show that the equation

$$x^{91} + x^{31} + 8x - 5 = 0$$

has exactly one real root.

OR*

Find the Taylor polynomial $T_3(x)$ at $c = 1$ for the function $f(x) = \frac{1}{x}$ then sketch the graph of f against T_3.

Solution

(a)

$$x^2 + 3y^2 = 1 \quad \text{therefore} \quad 2x + 6yy' = 0$$

$$\text{therefore} \quad y' = -\frac{x}{3y}$$

$$= 1$$

$$\Leftrightarrow \; x = -3y$$

i.e. Tangent lines with unit slope occur when

$$(-3y)^2 + 3y^2 = 1 \quad \text{i.e. } 12y^2 = 1$$

i.e.

$$y = \pm\frac{1}{\sqrt{12}}$$

therefore

$$x^2 = \frac{3}{4} \quad \text{i.e. } x = \pm\frac{\sqrt{3}}{2}$$

i.e. at $\left(\frac{\sqrt{3}}{2}, -\frac{1}{\sqrt{12}}\right)$ and $\left(-\frac{\sqrt{3}}{2}, \frac{1}{\sqrt{12}}\right)$

Equations of tangent lines:

$$y \pm \frac{1}{\sqrt{12}} = x \mp \frac{\sqrt{3}}{2}$$

(b)

$$f(x) = x^{91} + x^{31} + 8x - 5 = 0$$

Since f is continuous and $f(0) = -5$ and $f(1) = 5$, the equation has at least one root $a \in (0, 1)$ by the Intermediate Value Theorem. Suppose the equation has another root $b \in \mathbb{R}$ with $a < b$. Then

$$f(a) = f(b) = 0$$

and by Rolle's Theorem

$$f'(x) = 91x^{90} + 31x^{30} + 8 = 0 \quad \text{has a root in } (a, b).$$

But clearly, $f'(x) \geq 8 \ \forall \ x \in \mathbb{R}$.

Thus $f'(x)$ cannot have a root in (a, b).

This contradicts the original assumption that $\exists$ a second root b of the equation $f(x) = 0$.

Hence a is the only real root of the equation.

OR*

$$\begin{aligned} f(x) &= x^{-1} & f(1) &= 1 \\ f'(x) &= -x^{-2} & f'(1) &= -1 \\ f''(x) &= 2x^{-3} & f''(1) &= 2 \\ f^{(3)}(x) &= -6x^{-4} & f^{(3)}(1) &= -6 \end{aligned}$$

$$\begin{aligned} T_3(x) &= 1 - (x-1) + (x-1)^2 - (x-1)^3 \\ &= -x^3 + 4x^2 - 6x + 4 \end{aligned}$$

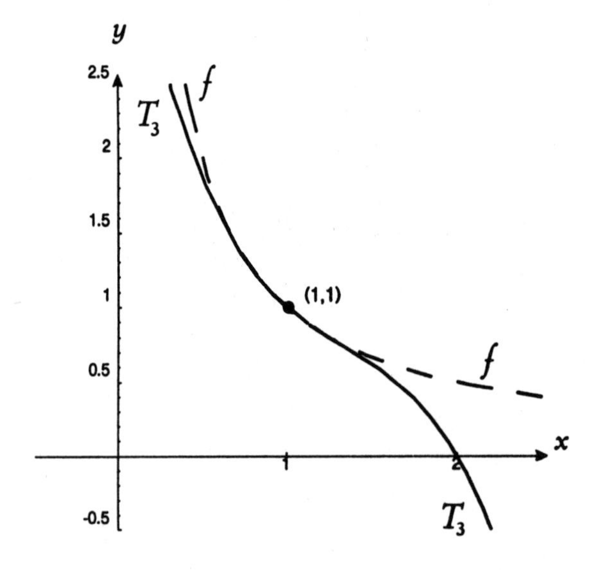

(10%) Q6. Use the Mean Value Theorem to show that if $f'(x) = 0 \ \forall\ x \in (a,b)$, then f is constant on (a,b).

OR*

Differentiate with respect to θ:
(i) $g(\theta) = [\arctan\sqrt{\theta}]^2$ (ii) $f(\theta) = e^{\theta}\ln\theta,\ \theta > 0$.

Solution

Let x_1, x_2 be any two numbers in (a,b) with $x_1 < x_2$. Since f is differentiable on (a,b) it must be differentiable on (x_1, x_2) and continuous on $[x_1, x_2]$. The Mean Value Theorem applied to f on $[x_1, x_2]$ gives a number c such that

$$f(x_2) - f(x_1) = f'(c)(x_2 - x_1), \quad x_1 < c < x_2 \qquad (*)$$

Since $f'(x) = 0,\ \forall x$, we have $f'(c) = 0$ so that (*) becomes

$$f(x_2) - f(x_1) = 0$$

or

$$f(x_2) = f(x_1)$$

i.e. f has the same value at *any* two numbers $x_1, x_2 \in (a,b)$ i.e. f is constant on (a,b).

OR*

(i) Let $g(\theta) = y = [\arctan\sqrt{\theta}]^2 = u^2,\ u = \arctan\sqrt{\theta}$

$$\begin{aligned} g'(\theta) = \frac{dy}{d\theta} &= \frac{dy}{du}\frac{du}{d\theta} \qquad \text{(chain rule)} \\ &= 2u\frac{d}{dw}\arctan w \cdot \frac{dw}{d\theta}, \quad w = \sqrt{\theta} \\ &= 2u\frac{1}{1+w^2}\cdot\frac{1}{2\sqrt{\theta}} \\ &= 2\arctan\sqrt{\theta}\cdot\frac{1}{2\sqrt{\theta}(1+\theta)} \end{aligned}$$

(ii)

$$\begin{aligned} f(\theta) &= e^{\theta}\ln\theta,\ \theta > 0 \\ f'(\theta) &= e^{\theta}\ln\theta + \frac{e^{\theta}}{\theta} \qquad \text{(product rule)} \\ &= e^{\theta}\left(\ln\theta + \frac{1}{\theta}\right) \end{aligned}$$

APPENDIX
Useful Formulae/Results From Calculus

Limits

$\lim_{\theta \to 0} \frac{\sin\theta}{\theta} = 1, \ \lim_{\theta \to 0} \cos\theta = 1, \ \lim_{\theta \to 0} \sin\theta = 0$

$\lim_{x \to \infty} \frac{1}{x^r} = 0, \ r > 0$ is rational.

$\lim_{x \to -\infty} \frac{1}{x^r} = 0, \ r > 0$ is rational such that x^r is defined $\forall \ x$.

Continuity

A function f is continuous at a point $x = a$ iff

$$\lim_{x \to a} f(x) = f(a).$$

More generally,

$$\lim_{x \to a^-} f(x) = \lim_{x \to a^+} f(x) = \lim_{x \to a} f(x) = f(a)$$

Differentiability

A function f is differentiable at a point $x = a$ iff the limit

$$\lim_{h \to 0} \frac{f(a+h) - f(a)}{h} = f'(a)$$

exists.

i.e.

$$\lim_{h \to 0+} \frac{f(a+h) - f(a)}{h} = \lim_{h \to 0^-} \frac{f(a+h) - f(a)}{h} = f'(a)$$

Fundamental Theorem of Calculus

$$\frac{d}{du} \int_a^u f(t)dt = f(u), \quad a \text{ is a constant} \qquad (*)$$

Notice that before we can apply this rule, we must have

(i) The upper limit of integration *identical* to the variable of differentiation.

(ii) A constant on the lower limit of integration.

This is why we often use the chain rule when dealing with derivatives of integrals.

For example,

$$\frac{d}{dx}\int_3^{x^2}\sqrt{1+t^2}dt = \frac{d}{dw}\int_3^{w}\sqrt{1+t^2}dt \cdot \frac{dw}{dx}$$

where $w = x^2$. Now we have the expression in a form suitable for (*) above.

Antidifferentiation

$$\int_{-a}^{a} f(x)dx = \begin{cases} 0, \text{ if } f \text{ is odd} \\ 2\int_0^a f(x)dx, \text{ if } f \text{ is even} \end{cases}$$

$$(f \text{ is odd } \Leftrightarrow f(-x) = -f(x)$$
$$(f \text{ is even } \Leftrightarrow f(-x) = f(x))$$

Don't forget the constant of integration when performing indefinite integration.

For example,

$$\int \cos x dx = \sin x + \mathbf{c}$$

Curve Sketching

Domain of f consists of values of $x \in \mathbb{R}$ for which f makes sense.

f increases when $f'(x) > 0$

f decreases when $f'(x) < 0$

Local extrema occur when $f'(x) = 0$

$f(x)$ is *concave up* when $f''(x) > 0$

$f(x)$ is *concave down* when $f''(x) < 0$

Points of Inflection occur when $f''(x)$ changes sign (unless that value of x has already been identified as a vertical asymptote).

Asymptotes of $y = f(x)$

Horizontal: Lines $y = a$ such that $\lim_{x\to\pm\infty} f(x) = a$.

Vertical: If $f(x) = \frac{g(x)}{h(x)}$ then vertical asymptotes are lines $x = x_i$ where x_i are the roots of $h(x) = 0$.

Slant: Lines $y = ax + b$ such that

$$\lim_{x\to\pm\infty} [f(x) - (ax + b)] = 0.$$

Exponentials and Logarithms*

$$\int e^x dx = e^x + c$$

$$\frac{d}{dx}e^x = e^x$$

$$\int \frac{1}{x} dx = \ln|x| + c$$

$$\frac{d}{dx}\ln x = \frac{1}{x}, \quad x > 0$$

* For students taking the early transcendentals option.